GOOD WOOD GUIDE

CORK CITY
LIBRARY

GOOD WOOD GUIDE

Albert Jackson & David Day

HarperCollins*Publishers*

GOOD WOOD GUIDE
Conceived, edited and designed at Inklink, Greenwich, London

Text: Albert Jackson and David Day

Design and art direction: Simon Jennings

Technical consultant: Peter B. Cornish, MDes (RCA), FSIAD

Editor: Ian Kearey

Illustrators: Robin Harris and David Day

Studio photography: Neil Waving and Ben Jennings

Indexer: Ian Kearey

First published in 1996 by
HarperCollins*Publishers*, London

9 8 7 6 5 4 3

For HarperCollins:
Editorial director: Polly Powell
Senior production manager: Bridget Scanlon

A CIP catalogue record is available from the British Library

ISBN 0 00 412997 0

Text set in Franklin Gothic Extra Condensed, Univers Condensed and Garamond Book Condensed by Inklink, London

Printed in Singapore

Jacket design: Simon Jennings
Jacket photograph: Neil Waving
Jacket illustrations: Robin Harris

CONTENTS

INTRODUCTION

Even the most experienced wood-worker would find it difficult to identify every species of timber; for the rest of us, recognizing more than commonly used woods can be fraught with confusion. In any case, choosing wood for a project requires far more than just appreciating its colour, figure, grain and texture. A knowledge of the wood's working characteristics and common uses – for instance, whether it will be strong enough for the job or how it takes a finish – can help prevent problems and make the work more pleasurable. In addition, man-made boards and veneers can be used, either in place of solid wood or for their own properties and effects.

It is also important that any new wood comes from a well-managed source, and we should all be aware of which species are at risk; this way, informed choices can be made and steps taken to ensure continued supplies of this beautiful material.

ACKNOWLEDGEMENTS

The authors would like to thank the following for the supply of reference material:
American Hardwood Export Council, London
American Plywood Association, London
Australian Particleboard Institute Inc., NSW, Australia
Stuart Batty, Derbys
Better Built Corporation, Wilmington, MA, USA
Blount UK Ltd, Tewkesbury, Glos
John Boddy Timber Ltd, Boroughbridge, N. Yorks
Buckinghamshire College, High Wycombe, Bucks
Council of Forest Industries of Canada, West Byfleet, Surrey
Craft Supplies, Buxton, Derbys
Karl Danzer Furnierwerke, Reutlingen, Germany
Department of the Environment, Bristol
English Nature, Peterborough, Cambs
Fastnet Products Ltd, Chard, Somerset
Finnish Forest Industries Federation, Helsinki, Finland
Walter Fischer, Kassel, Germany
Fitchett & Woolacott, Nottingham
Forest Stewardship Council, Oaxaca, Mexico
Forests Forever, London
Friends of the Earth, London
Furniture Industry Research Association, Stevenage, Herts
Granberg International, Richmond, CA, USA
Hardwood Plywood and Veneer Association, Reston, VA, USA
Legno Ltd, London
Louisiana-Pacific Corp., Portland, OR, USA
Malaysian Timber Council, London
Marlwood Ltd, Edenbridge, Kent
Metsäkuva-Arkisto KY, Helsinki, Finland
Milland Fine Timber Ltd, Liphook, Hants
Theodor Nagel GmbH & Co., Hamburg, Germany
Oxford Forestry Institute, Oxford
Plywood Association of Australia Ltd, Newstead, QLD, Australia
Royal Botanic Gardens, Kew, Surrey
Schauman Wood Oy, Helsinki, Finland
Southeastern Lumber Manufacturers Association, Forest Park, GA, USA
Timber Development Association (NSW) Ltd, NSW, Australia
Timber Research and Development Association, High Wycombe, Bucks
Timber Trade Federation, London
Union Veneers plc, London
US Forest Products Laboratory, Madison, WI, USA
Wagner Europe Ltd, Folkestone, Kent
WMS Consulting Ltd, Folkestone, Kent

The producers of the book are grateful to the following for the supply and preparation of woods, veneers, man-made boards and tools:
C.F. Anderson & Son Ltd, London
Annandale Timber & Moulding Co. Pty Ltd, NSW, Australia
Art Veneer Co, Mildenhall, Suffolk
John Boddy Timber Ltd, Boroughbridge, N. Yorks
Jim Cummins, Woodstock, NY, USA
Desfab, Beckenham, Kent
Egger (UK) Ltd, Hexham, Northumberland
FIDOR, Feltham, Middx
General Woodworking Supplies, London
Highland Forest Products plc, Highland Region
E. Jones & Son, Erith, Kent
Limehouse Timber, Dunmow, Essex
Ravensbourne College of Design & Communication, Chislehurst, Kent
Seaboard International, London
Fred Spalding, Bromley, Kent

Technical consultant
Peter B. Cornish is Head of Department, Faculty of Design, at Buckinghamshire College, High Wycombe, Bucks.

Photography
The studio photographs for this book were taken by Neil Waving, *with the following exceptions:*
Ben Jennings, pages 31, 37 (B), 93 (R), 99 (BR)

The producers are also indebted to the following for the use of photographs:
Gavin Jordan, pages 10, 16
Karl Danzer Furnierwerke, pages 13 (BR), 27 (TL), 84 (BL)
Simo Hannelius, pages 18, 19, 42, 106
Southeastern Lumber Manufacturers Association (Jim Lee), pages 20, 22, 86 (BL)
Practical Woodworking, page 27 (BL)
Wagner Europe, page 27 (BR)
Council of Forest Industries of Canada, pages 28, 43 (BL)
Buckinghamshire College, pages 34 (TL), 108 (L)
International Festival of the Sea (Peter Chesworth), page 36 (B)
John Hunnex, page 36 (T)
Stewart Linford Furniture (Theo Bergström), page 37 (TL)
Roger Bamber, pages 40–1
Malaysian Timber Council, pages 54, 55 (BL)
Schauman Wood Oy, page 85

Designer-makers
Philip Larner, pages 34 (TL), 108 (L)
John Hunnex, page 36 (T)
Stewart Linford, page 37 (TL)
Derek Pearce, page 37 (TR)
Mike Scott, page 37 (CR)

Key to credits
T = top, B = bottom, L = left, R = right, TL = top left, TC = top center, TR = top right, CL = center left, C = center, CR = center right, BL = bottom left, BC = bottom center, BR = bottom right

CHAPTER 1

THE RAW MATERIAL

Though some woodworkers can't see the trees for the wood, it is useful to understand something of the physical make-up of trees, and of how they grow. By doing so, you will be able to appreciate how the properties of each species of wood, and of different parts of the same tree, affect the ways this unique material can be worked and finished.

THE ORIGINS OF WOOD

Trees, whether growing in forests or standing alone, not only help control our climate but also provide habitats for a vast number of plants and living creatures. Tree derivatives range from natural foodstuffs through to extracts used in manufacturing products such as resins, rubber and pharmaceuticals. When cut down and converted into wood, trees provide an infinitely adaptable and universally useful material.

What makes a tree?

Botanically, trees belong to the *Spermatophyta* – a division of seed-bearing plants which is sub-divided into *Gymnospermae* and *Angiospermae*. The former are needle-leaved coniferous trees, known as softwoods, and the latter are broadleaved trees that may be deciduous or evergreen; these are known as hardwoods. All trees are perennials, which means they continue their growth for at least three years.

The main stem of a typical tree is known as a bole or trunk, and carries a crown of leaf-bearing branches. A root system both anchors the tree in the ground and absorbs water and minerals to sustain it. The outer layer of the trunk acts as a conduit to carry sap from the roots to the leaves.

Mature montane coniferous forest

Nutrients and photosynthesis

Trees take in carbon dioxide from the air through pores in the leaves called stomata, and evaporation from the leaves draws the sap through minute cells (see below). When the green pigment present in leaves absorbs energy from sunlight, organic compounds are made from carbon dioxide and water. This reaction, called photosynthesis, produces the nutrients on which a tree lives, and at the same time gives off oxygen into the atmosphere. The nutrient produced by the leaves is dispersed down through the tree to the growing parts, and is also stored by particular cells.

Although it is often thought that wood 'breathes' and needs to be nourished as part of its maintenance, once a tree is felled, it dies. Any subsequent swelling or shrinking is simply a reaction of the wood to its environment as it absorbs and exudes moisture, in a similar way to a sponge. Wood finishes such as waxes and oils enhance and protect the surface, and to some extent help stabilize movement, but they do not 'feed' the wood.

Cellular structure

A mass of cellulose tubular cells bond together with lignin, an organic chemical, to form the structure of wood. These cells provide support for the tree, circulation of sap and food storage. They vary in size, shape and distribution, but are generally long and thin and run longitudinally with the main axis of the tree's trunk or branches. The orientation produces characteristics relating to the direction of grain, and the varying size and distribution of cells between species produce the character of wood textures, from fine to coarse.

Identifying wood

Examination of cells enables the identification of cut wood as being a softwood or hardwood. The simple cell structure of softwoods is composed mainly of tracheid cells which provide initial sap conduction and physical support. They form regular radiating rows and make up the main body of the tree.

Hardwoods have fewer tracheids than softwoods; instead, they have vessels or pores which conduct sap, and fibres that give support.

Carbon dioxide
Drawn in from the air through leaves.
Oxygen
Given out into the air by living trees.
Leaf-bearing branches
Leaves produce nutrients that feed the tree by photosynthesis.
Trunk
The trunk supports the leaf-bearing branches and is the main source of useful wood.
Gymnosperms – needle-leaved trees
Angiosperms – broadleaved trees
Root system
Roots anchor the tree and absorb moisture and minerals from the soil.

Ph/copy.

HOW TREES GROW

A thin layer of living cells between the bark and the wood, called the cambium, sub-divides every year to form new wood on the inner side and phloem or bast on the outside. As the inner girth of the tree increases, the old bark splits and new bark is formed by the bast. Cambial cells are weak and thin-walled; in the growing season, when they are moisture-laden, the bark can be easily peeled. In winter months, the cells stiffen and bind the bark firmly. The new wood cells on the inside develop into two specialized types: living cells which store food for the tree, and non-living cells which conduct sap up the tree and provide support for it. These two types make up the sapwood layer.

Each year, a new ring of sapwood is built up on the outside of the previous year's growth. At the same time, the oldest sapwood nearer the centre is no longer used to conduct water; it is chemically converted into the heartwood that forms the structural spine of the tree. The area of heartwood increases annually, while the sapwood remains at around the same thickness during the tree's life.

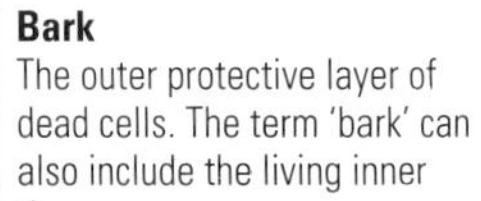

Bark
The outer protective layer of dead cells. The term 'bark' can also include the living inner tissue.

Bast or phloem
The inner bark tissue that conducts synthesized food.

European oak
Quercus petraea
The photograph shows a cross section of a European-oak trunk (see page 78).

Cambium layer
The thin layer of living cell tissue that forms new wood and bark.

Sapwood
The new wood, the cells of which conduct or store nutrients.

Annual-growth ring
The layer of wood formed in one growing period, made up of large earlywood and small latewood cells.

Ray cells
Radiating sheets of cells that conduct nutrients horizontally; also called 'medullary rays'.

Heartwood
The mature wood that forms the tree's spine.

Pith
The central core of cells. This can be weak and often suffers from fungal and insect attack.

Ray cells

Ray cells, or medullary cells, radiate from the centre of the tree. They carry and store nutrients horizontally through the sapwood, in the same way as the cells that follow the axis of the trunk. The flat vertical bands formed by ray cells can hardly be detected in softwoods; in some hardwoods such as oak, particularly when it is quarter-sawn, the ray cells are plainly visible.

Sapwood

Sapwood can usually be recognized by its lighter colour, which contrasts with the darker heartwood. However, this difference is less distinct on light-coloured woods, particularly softwoods. Because sapwood cells are relatively thin-walled and porous, they tend to give up moisture quickly and shrink more than the denser heartwood. Conversely, this porosity means that sapwood can readily absorb stains and preservatives.

For the woodworker, sapwood is inferior to heartwood; furniture makers usually cut it to waste. It is not very resistant to fungal decay, and the carbohydrates stored in some cells are also liable to insect attack.

Heartwood

The dead sapwood cells that form heartwood have no further part in the tree's growth, and can become blocked with organic material. Hardwoods with blocked cells – white oak, for instance – are impervious and much better suited to tasks such as tight cooperage than woods like red oak, which have open heartwood cells and are thus relatively porous.

The chemical substances that cause the dead cell walls to change colour, sometimes deeply in the case of hardwoods, are called extractives. They also provide some resistance to insect and fungal attack.

Annual rings

The distinct banding made by earlywood and latewood corresponds to one season's growth and enables the age of a felled tree, and the climatic conditions through which it has grown, to be determined. In the simplest example, wide annual rings indicate good growing conditions, narrow ones poor or drought conditions, but study of the annual rings can tell the history of the tree's growth in detail.

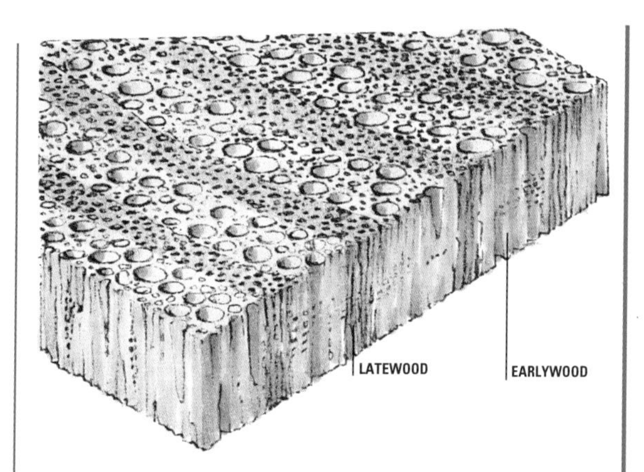

Earlywood

Earlywood, or springwood, is the rapid part of the annual-growth ring that is laid down in spring, at the early part of the growing season. In softwood, thin-walled tracheid cells form the bulk of the earlywood and facilitate the rapid conduction of sap. In hardwood, open tube-like vessels perform the same function. Earlywood can usually be recognized as the wider band or paler-coloured wood in each growth ring.

Latewood

Latewood, or summerwood, grows more slowly in the summertime, and produces thicker-walled cells. Their slower growth creates harder and usually darker wood, which is less able to conduct sap but provides support to the tree.

Young hardwood forest

CONSERVATION AND TREES

NB.

Tree products are a natural resource that is widely and increasingly used in all our lives; trees also play a vital part in regulating our environment. Increasing ecological awareness has highlighted the plight of forests endangered by overcutting and pollution; other, more efficient, energy sources must be developed and carbon-dioxide emissions and other pollutants controlled.

Environmental threats

Carbon dioxide, a by-product of burning fossil fuels, makes up part of the earth's atmosphere. Living trees absorb this gas (see page 10), which helps maintain the natural balance of the atmosphere. However, the level of carbon dioxide is rising faster than can be naturally absorbed, leading to the greenhouse effect, where carbon dioxide and other gases trap the earth's radiated heat, leading to global warming.

In the southern hemisphere, deliberate burning of the Amazonian rainforests, to clear land for farming and cattle ranching, not only reduces the stock of virgin forest but also contributes to the greenhouse effect. Polluted air from industries in the northern hemisphere produces acid rain that is killing trees and entire forests.

CITES REGULATIONS

At the time of writing, the only international conservation regulations to be implemented are those listed by CITES – the Convention on International Trade in Endangered Species of Wild Flora and Fauna. The CITES regulations have three appendices or grades. At regular, biennial meetings, all species on the three appendices are reviewed, and additions and deletions from each appendix are made.

Appendix I lists species threatened with extinction, in which all commercial imports, exports and sales, including seeds and manufactured products (new or antique), are prohibited.

Appendix II includes species which may become threatened if trade in them is not controlled and monitored; any export must be accompanied by an Export Certificate issued by a government authority of the exporting country, after investigation that the specimen was legally obtained and that export would not be detrimental to the survival of that species. Importers must obtain an Import Certificate.

Species in Appendix III are noted as being threatened or at risk by any country to which the species is indigenous. The listing provides that country with a measure of control over the amount exported, while the importer must still have a CITES Import Certificate.

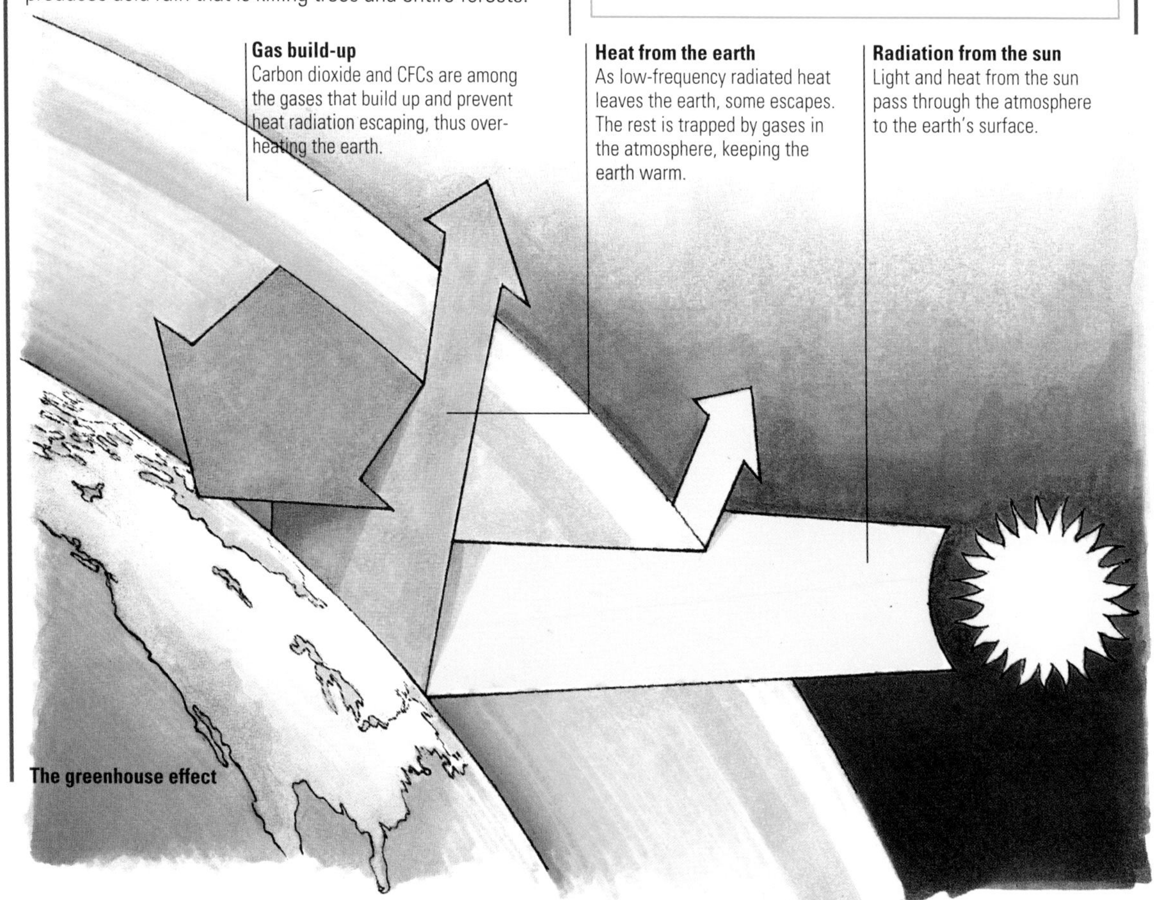

The greenhouse effect

Growing concerns

Depending on whom you speak to, the survival of tropical hardwoods and even some South-American softwoods is either in crisis and at the point of collapse, or is being responsibly managed. The reality is somewhere in between these statements.

Environmental groups have drawn the Western world's attention to mass defoliation of tropical forests and rainforests. However, a complete ban on imported tropical hardwoods, as suggested by some, would damage the timber trade and deprive developing countries of revenue. In addition, more felled timber is destroyed by local burning than is logged or exported. Multinational mining and dam-construction projects also contribute to the problem, as does the pulping industry, by stripping mixed virgin forest for monoculture reforestation.

The timber trade

Trees are a renewable resource, and responsible programmes could make it possible to ensure a continued supply of tropical hardwoods. There is increasing pressure on timber producers, suppliers and users to trade in and work only those woods that come from certifiably managed sources.

Organisations such as the Forest Stewardship Council and Forests Forever include timber traders and suppliers among their ranks, and aim to increase wood-users' responsibility for, and knowledge about, their material. Both these groups are in the process of devising methods of certifying tropical, temperate and boreal woods according to the way they are grown and managed by foresters and timber companies.

Using alternatives

For the woodworker, CITES regulations mean that certain exotic timbers will either become much less available or be found only among old stock, or as reclaimed timber. It is therefore worth considering the use of alternative timbers and those produced in temperate zones (see page 17). Reputable timber suppliers will be able to advise you about alternative woods and on the availability of tropical hardwoods from reputable sources, and can help you to 'act locally, think globally'.

ENDANGERED TREES

Lignum vitae **Afrormosia** **Brazilian rosewood**

Lignum vitae – *Guaiacum officinale* (see page 70) – and **afrormosia** – *Pericopsis elata* (see page 74) – are listed under CITES Appendix II, which controls the supply of logs, sawn wood and veneers. **Brazilian rosewood** – *Dalbergia nigra* (below)– is threatened with extinction and listed under CITES Appendix I. This beautifully coloured and figured wood, once widely used for fine furniture, turnery and carving, is no longer commercially available.

Brazilian rosewood

Temperate hardwoods

Timbers of the temperate forests of North America and Europe are already produced by sustainable methods. In the United States, the Multiple-Use Sustained-Yield Act requires that trees harvested from Federal land do not exceed the annual growth. It recognizes that public forests also provide wildlife habitats and control watershed and soil erosion, as well as being used for recreational purposes. A 30-year policy of continual regeneration has produced around 50 per cent more hardwood than has been used in the same time.

Most commercial hardwoods are from second- third- or fourth-cut forests that are managed on a rotational basis. The remaining virgin forests are now protected from commercial logging, and no old-growth timber is felled.

Temperate hardwoods may not offer such a wide choice of colour as 'exotic' tropical species, but wood stains can be used to modify colour. Refer to the chart opposite for possible alternatives.

Tropical hardwoods

When looking at a finished piece of woodwork, it is easy to see why tropical hardwoods, with their diversity of figure and colour, are so sought-after. Unfortunately, some species have been over-harvested, and we are heading for a situation where these materials will no longer be readily available.

For example, the rich red colouring of mahogany made it a favourite for furniture-making in Victorian times, and it remains so today. It has become one of the main tropical timbers traded across the developed world and used in vast quantities in the building industry and domestic furnishing.

Types of mahogany

Mahogany has become the generic name for a number of similar woods. Those from South America are described as 'true' mahogany (*Swietenia* spp.), and are usually known as Cuban, Honduras, Spanish or Brazilian mahogany. Of these, Brazilian (*Swietenia macrophylla*) is the most common.

African mahogany is from the botanical group *Khaya*. To meet growing demand, woods from the genus *Etandrophragma*, usually called sapele or utile, are also traded as mahogany, as they have similar properties. To add to the confusion, red lauan (*Shorea negrosensis*) is sometimes erroneously referred to as Philippine mahogany.

Swietenia **(mahogany)**

A species in decline

The *Swietenia* species from South America is among other species in decline due to exploitation, which is causing serious environmental problems. Heavy machinery and the dirt roads it makes have caused a disproportionate amount of damage to the forest, compared with the quantity of timber extracted. Unfortunately, mahogany does not easily re-seed itself in logged forest areas, resulting in little natural regeneration. The lands of the indigenous people have been plundered illegally by logging companies extracting the trees; all too often, wholesale clearing of the forest, usually by fire, is followed by agricultural development by migrants. Responsible importers deal only with certificated material, and in addition pay a levy on shipments; this levy is used to promote improved forest management in the country of origin.

'Slash-and-burn' forest clearance by fire

INSTANT GUIDE TO COMMON USES

The chart below gives an instant reference to the main uses of all the woods included in this book.

	Building construction	Joinery – exterior	Joinery – interior	Doors	Flooring	Furniture/cabinetmaking	Turnery	Carving/pattern-making	Musical instruments	Sports equipment	Boxes/crates	Boatbuilding	Tool handles/implements
SOFTWOODS													
Silver fir, *Abies alba*	◊		•		•						•		
Queensland kauri, *Agathis* spp.			•			•							
Parana pine, *Araucaria angustifolia*	◊		•			•	•						
Hoop pine, *Araucaria cunninghamii*	◊		•		•	•	•	•			•		
Cedar of Lebanon, *Cedrus libani*	◊	•	•	•		•							
Yellow cedar, *Chamaecyparis nootkatensis*	◊	•	•	•	•	•						•	•
Rimu, *Dacrydium cupressinum*	◊	•	•		•	•							
Larch, *Larix decidua*	□	•	•		•	•						•	•
Norway spruce, *Picea abies*	◊	•	•		•	•			•		•		
Sitka spruce, *Picea sitchensis*	◊		•			•	•		•	•	•	•	
Sugar pine, *Pinus lambertiana*	◊		•	•							•	•	
Western white pine, *Pinus monticola*	◊		•	•	•	•		•				•	
Ponderosa pine, *Pinus ponderosa*	◊		•	•	•	•	•	•			•	•	
Yellow pine, *Pinus strobus*	◊		•	•	•	•		•	•			•	
European redwood, *Pinus sylvestris*	□	•	•	•	•	•	•		•		•	•	•
Douglas fir, *Pseudotsuga menziesii*	□	•	•	•	•	•					•	•	
Sequoia, *Sequoia sempervirens*	◊	•	•						•		•	•	
Yew, *Taxus baccata*			•			•	•	•		•			
Western red cedar, *Thuja plicata*	◊	•	•			•					•	•	
Western hemlock, *Tsuga heterophylla*	□	•	•	•	•	•							
HARDWOODS													
Australian blackwood, *Acacia melanoxylon*			•			•	•			•			•
European sycamore, *Acer pseudoplatanus*			•		•	•	•	•	•	•			•
Soft maple, *Acer rubrum*			•	•	•	•	•		•	•			
Hard maple, *Acer saccharum*			•	•	•	•	•		•	•			•
Red alder, *Alnus rubra*						•	•	•					
Gonçalo alves, *Astronium fraxinifolium*		•				•	•						
Yellow birch, *Betula alleghaniensis*			•	•	•	•	•		•				
Paper birch, *Betula papyrifera*							•		•		•		•
Boxwood, *Buxus sempervirens*						•	•	•	•	•			•
Silky oak, *Cardwellia sublimis*	Δ		•		•	•							
Pecan hickory, *Carya illinoensis*						•	•			•			•
American chestnut, *Castanea dentata*		•	•			•					•		
Sweet chestnut, *Castanea sativa*	◊	•	•			•	•						
Blackbean, *Castanospermum australe*	Δ		•			•	•	•		•			•
Satinwood, *Chloroxylon swietenia*	Δ	•	•			•	•	•					
Kingwood, *Dalbergia cearensis*						•	•						
Indian rosewood, *Dalbergia latifolia*			•		•	•	•		•			•	
Cocobolo, *Dalbergia retusa*							•	•					•
Ebony, *Diospyros ebenum*						•	•	•	•	•			•
Jelutong, *Dyera costulata*			•	•				•					
Queensland walnut, *Endiandra palmerstonii*			•		•	•							
Utile, *Entandrophragma utile*	◊	•	•	•	•	•			•	•		•	
Jarrah, *Eucalyptus marginata*	Δ	•	•		•	•	•					•	•
American beech, *Fagus grandifolia*	◊		•		•	•	•		•	•		•	•
European beech, *Fagus sylvatica*	□		•		•	•	•		•	•		•	•
American white ash, *Fraxinus americana*	◊		•			•				•		•	•
European ash, *Fraxinus excelsior*	◊		•			•	•			•		•	•
Ramin, *Gonystylus macrophyllum*			•		•	•	•	•					•
Lignum vitae, *Guaiacum officinale*							•			•		•	•
Bubinga, *Guibourtia demeusei*		•	•			•	•						•
Brazilwood, *Guilandina echinata*	Δ	•	•		•	•	•		•	•			
Butternut, *Juglans cinerea*			•			•		•			•	•	
American walnut, *Juglans nigra*			•	•		•	•	•	•	•		•	
European walnut, *Juglans regia*			•	•		•	•	•		•			
American whitewood, *Liriodendron tulipifera*	◊		•	•		•		•				•	
Balsa, *Ochroma lagopus*						•						•	
Purpleheart, *Peltogyne* spp.	Δ	•	•		•	•	•			•		•	•
Afrormosia, *Pericopsis elata*	□	•	•	•	•	•						•	•
European plane, *Platanus acerifolia*	◊		•	•		•	•					•	
American sycamore, *Platanus occidentalis*	◊		•	•		•	•						
American cherry, *Prunus serotina*			•			•	•	•	•			•	
African padauk, *Pterocarpus soyauxii*	Δ	•	•		•	•	•	•		•		•	•
American white oak, *Quercus alba*	Δ	•	•	•	•	•	•	•				•	•
Japanese oak, *Quercus mongolica*	□	•	•	•	•	•	•	•				•	•
European oak, *Quercus robur/petraea*	□	•	•	•	•	•	•	•				•	•
American red oak, *Quercus rubra*	□		•		•	•	•					•	
Red lauan, *Shorea negrosensis*		•	•	•		•		•	•	•	•	•	
Brazilian mahogany, *Swietenia macrophylla*	Δ	•	•	•	•	•	•	•	•	•	•	•	
Teak, *Tectona grandis*	□	•	•	•	•	•	•	•				•	
Basswood, *Tilia americana*			•			•	•	•	•		•		•
Lime, *Tilia vulgaris*			•				•	•	•	•	•		•
Obeche, *Triplochiton scleroxylon*			•			•			•	•			
American white elm, *Ulmus americana*	Δ		•		•	•	•	•			•	•	•
Dutch/English elm, *Ulmus hollandica/procera*	Δ		•		•	•	•	•				•	•

Building-construction key: Δ = Heavy construction ◊ = Light construction □ = Light or heavy

FORESTRY AND CULTIVATION

As a source of raw material, trees can be regarded as victims of their own success. In the past, forests appeared to provide an inexhaustible supply of wood, and trees were cut down with little or no thought for the future. The virgin forests of Europe were exhausted long ago, and it was only relatively recently that legislation prevented similar depletion in North America. Today, however, developing countries feel it necessary to exploit their natural forest resources for short-term financial gain.

Ecosystems

The forest ecosystem is a highly complex, symbiotic, interrelated life-support system for a diverse range of flora and fauna. Each plant species contributes to the life cycle of the others.

A natural forest takes hundreds of years to mature. Primitive vegetation, in the form of lichens and mosses, first colonizes bare rock surfaces. In time, these plants help build up a soil layer, which enables a higher order of flowering plants to become established. These in turn are joined by larger shrub plants and then by young trees that become a mature forest.

Left alone, the forest forms a stable environment, but should the balance be altered by climate or natural hazards that affect localized areas, such as fire, the plants best suited to the changed conditions will become dominant. However, if the ecosystem is radically altered by outside agencies, there is less chance of the forest remaining self-sustaining.

Silviculture

The cultivation of man-made forests is called silviculture. Modern forestry makes use of scientific research to create and control the development of productive forests that can sustain a successful ecosystem in a relatively short timescale. This is achieved through various methods: careful breeding, propagation, selection and planting of light-seeking or shade-tolerant tree species with a preference for the local soil; the removal of weak trees in favour of more robust specimens; clearing and replanting sections of the forest with alternative species; encouraging natural seeding where appropriate; controlling pests and diseases, and continually monitoring the overall health and condition of the forest.

Planting silver birch using a round hoe

Man-made forests

The main purpose for creating a man-made forest is to establish a renewable resource for the production of commercial timber. Other factors, such as the effect on local climate, soil erosion or flooding, and recreational uses, are equally important considerations.

In the Western world, new forests are usually formed from fast-growing coniferous trees, which are more adaptable to harsh conditions than slow-growing hardwoods and can establish themselves quickly as 'pioneers'. These monoculture forests provide a useful commercial crop in a relatively short time, for use in major industries such as building construction, pulp for paper-making, and the production of man-made boards.

Broadleaved hardwood trees may be included at the outset, but slow-growing species are usually introduced once the initial plantation has been established, and at such a time when the resulting ecosystem will support a diversity of species.

Establishing the forest

Forest farming has become highly mechanized for maximum efficiency in meeting long- and short-term economic targets. Young trees are now bred from seed in nurseries and planted out in regular rows. The original planting is as much as ten times the expected output of mature trees, allowing for failure of young trees due to natural causes, as well as for intervention by foresters in the selective management of the trees. Where nature was once left alone, aircraft now cover the dense forest with fertilizers, insecticides or fungicides, to encourage growth and minimize losses.

Harvesting the forest

Forests are typically harvested on a crop-rotation basis. In this system, part of the forest is entirely cleared and replanted, and the remaining forest provides protection for the new plantation (which could be an alternative species) without radically altering the ecological balance. Other sections are harvested on reaching maturity. Where a large area is cleared, selected trees may be left standing to provide seed for the next generation. Once the new forest is established, mature trees are cut and young trees thinned out, enabling selected trees to develop into a mature forest.

Forest by-products

In addition to wood and pulp, trees provide a diverse range of materials, extracted from the living tree or from the wood. These include bark for bottle corks and leather tanning, and liquid extracts used in making rosin, turpentine, oils and tar, rubber, gums, vitamins and waxes, as well as edible syrups, fruits, nuts and fibres. Established forests also provide a habitat for many medicinal plants. It is therefore vital that the true value of the forest is recognized and appreciated, and that appropriate capital is invested to develop well-managed natural forests, in order to preserve these unique renewable resources.

Man-made silver-birch forest

LOGGING

In past centuries, rural communities were able to exploit abundant natural forests and cut down stands of trees as required for house-building, fuel, ground clearance for agriculture, and for establishing boundaries. Today, most forests are managed on a commercial basis by the forest industries, a combination of companies involved in the manufacture and marketing of timber, and government organizations.

Seasonal logging

Logging is to some extent a seasonal business, governed by the geographical location of the forest. In the coastal regions of North America, for example, logging can continue virtually all year round, except for periods of drought, when there is a high risk of fire, or heavy snowfalls in winter. In forests situated in the interior, logging is mainly conducted from the winter into late spring, because the sap level is low during this period and the wood is relatively light in weight.

Commercial logging: felling a tree with a chainsaw

Using a chainsaw
Chainsaws are dangerous tools if not used properly, and must always be handled and maintained according to the manufacturer's instructions. A hard hat, safety goggles, ear protectors, tough work boots and strong gloves are needed as basic safety equipment. Protective garments are also recommended, to protect limbs against an accidental slip with the saw. Chainsaw training courses are available.

Commercial-logging methods

Trees were traditionally felled by teams of men using axes and hand saws, and the logs were then hauled by oxen or steam-driven machines. The logging industry still uses a large labour force, but powered chainsaws and mobile, mechanical tree shears are now used to cut trees. The logs are extracted from the forest by heavy, diesel-powered lifting equipment, or are hauled or lifted out on cables that are set up to move them from the felling area to the collection site. The logs are then loaded onto trucks for transportation, or are floated as log rafts, or 'booms', to the sawmills.

Felling your own trees

For those woodworkers who own their own plot of woodland that can be harvested, or who have access to private land, it is possible to fell trees using basic equipment. However, cutting down a tree should not be considered lightly, as it is physically demanding and dangerous work that requires careful preparation. Felling trees may not be possible if the trees are situated in a conservation area or if a local authority has put a protection order on them.

Tools and equipment

A small tree can be cut with a hand saw, but a chainsaw can both make the job easier and handle larger work. Chainsaws can be purchased from most tool suppliers or hired for occasional use. Electric- and petrol-powered chainsaws are available; the latter offer greater flexibility of movement. A medium-size saw should be sufficient for domestic use, and a hatchet or pruning saw may be needed for clearing ground-level shoots and branches.

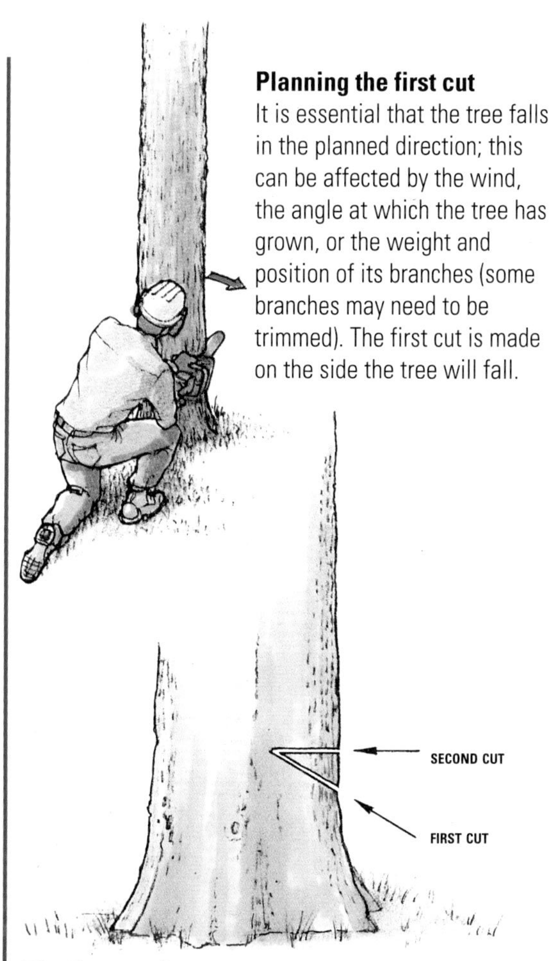

Planning the first cut

It is essential that the tree falls in the planned direction; this can be affected by the wind, the angle at which the tree has grown, or the weight and position of its branches (some branches may need to be trimmed). The first cut is made on the side the tree will fall.

The first notch

A wedge of wood is removed, to encourage the tree to fall as planned. The first cut is made at an upward angle of about 45 degrees and to a depth of around one-third the tree's diameter. The second cut is made parallel to the ground, to meet the first cut and to form the notch.

SITE SAFETY

- *Never attempt to fell a tree without thorough preparation of the tools, equipment and felling site.*
- *Do not work or utilize electrical appliances in wet conditions.*
- *Keep fuel oils well clear of the work area.*
- *Secure pets, and keep children and onlookers away from the work area.*
- *Ensure the site is clear of all loose material underfoot, and trim back surrounding vegetation to give a clear path of retreat.*
- *Make sure the tree will fall into a clear space.*
- *Take great care when felling a rotten tree, because the fall direction is less predictable than that of a sound one.*

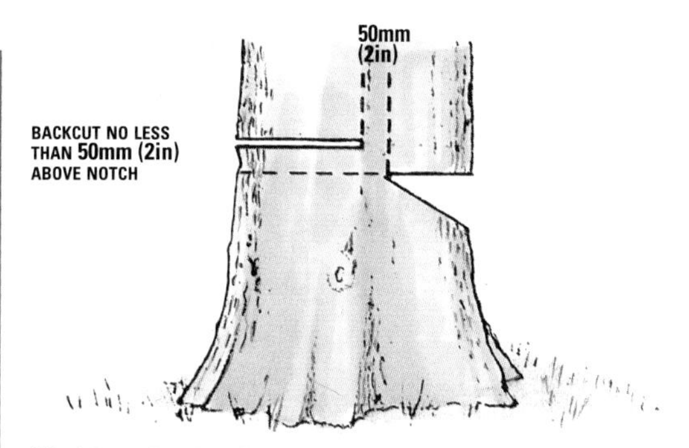

Making the backcut

The backcut fells the tree. It is made on the reverse side from the notch, not less than 50mm (2in) above it, and must stop about 50mm (2in) short of the notch to leave a section of uncut wood. This 'hinge' controls the direction of fall and helps prevent 'kickback' of the trunk. Larger trees may need wedges driven in the backcut, to prevent the saw blade being pinched.

Avoiding the falling tree

As soon as the tree starts to topple, the chainsaw must be stopped and put down quickly, in order to move away from the tree.

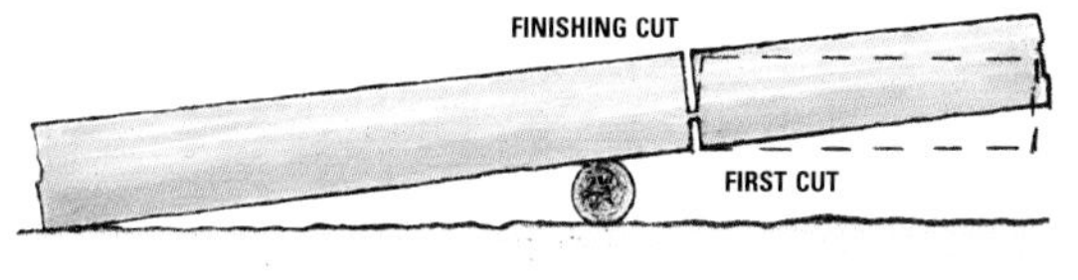

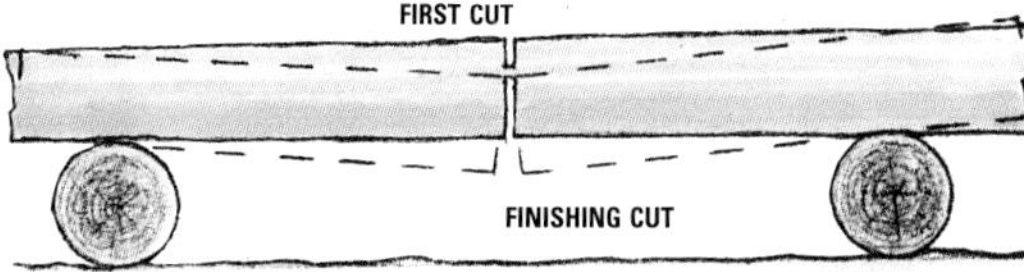

Preparing the logs

The felled tree is trimmed of its limbs and branches using a chainsaw, taking care that the limbs do not spring back as they are cut. The trimmed trunk is then cut into logs or 'bucked' into manageable lengths. To avoid the log pinching the saw, two meeting cuts are made – one from above and one from below – using both edges of the blade. The first cut penetrates about one-third of the log's diameter, and is made on the side that will tend to close on the saw; this depends on how the log is supported. The finishing cut is made to meet the first cut.

CONVERTING WOOD

Although it can take many years for a tree to grow to a commercially viable size, modern forestry methods can cut down, top and de-bark a straight-growing tree, such as pine, in a matter of minutes. In the same way, the laborious task of sawing logs into boards or beams by hand in a saw-pit has been superseded; today, milling is a highly mechanized process, where logs are converted into sawn timber by computer-controlled band-saws or circular saws.

Converting a log into timber at a sawmill

Trunks and branches

The trunk of the tree supplies most usable commercial wood. Although larger limbs can be cut into logs, asymmetric growth rings in branches or slanted trunks usually produce 'reaction' wood, which is unstable and warps and splits easily. In softwoods, the annual-ring growth is mainly on the underside of the branch, and produces 'compression wood'. In hardwoods, it grows mainly on the upper side, and is called 'tension wood'.

Good-quality felled trees are cut into logs or butts and transported to local sawmills for conversion into rough-sawn timber. Top-quality hardwood logs that have large, even boles fetch high prices and are usually converted into veneer (see page 88). Tree trimmings, sub-grade wood and forest thinnings are generally used for manufactured boards and paper products.

TYPES OF CUTTING

Plain-sawn boards in Britain and Europe have growth rings meeting the face of the board at an angle of less than 45 degrees. In quarter-sawn boards, the angle of growth rings to the face of the board is greater than 45 degrees.

Plain-sawn boards in North America have growth rings meeting the face at an angle of less than 30 degrees. Boards where the rings meet between 30 and 60 degrees are called rift-sawn; these boards display straight figure with some ray-cell patterning, and are sometimes referred to as comb grain.

True quarter-sawn boards are cut radially, with the annual rings perpendicular to the board's face, but in practice boards with the rings at an angle of not less than 60 degrees are classified as quarter-sawn.

Milling

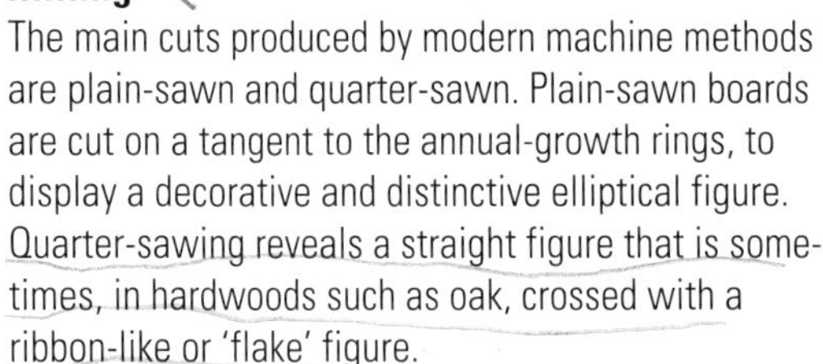

The main cuts produced by modern machine methods are plain-sawn and quarter-sawn. Plain-sawn boards are cut on a tangent to the annual-growth rings, to display a decorative and distinctive elliptical figure. Quarter-sawing reveals a straight figure that is sometimes, in hardwoods such as oak, crossed with a ribbon-like or 'flake' figure.

Different terms are used for boards within the two categories. Plain-sawn timber is also known as flat-sawn, flat-grain, or slash-sawn timber. Quarter-sawn timber includes rift-sawn, comb-grain, edge-grain and vertical-grain.

Types of cut, from top to bottom: Plain-sawn; rift-sawn; quarter-sawn.

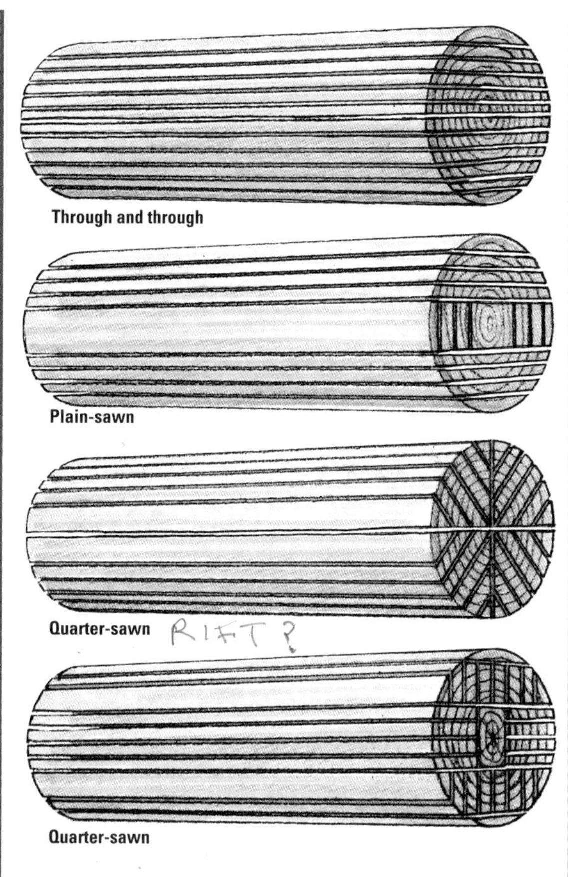

Cutting through and through

The stability and figure of wood are determined by the relationship of the saw-cut plane to the annual-growth rings. The most economical method of converting a log is to cut it 'through and through'; in this process, parallel cuts are made through the length of the log to produce plain-sawn, rift-sawn and a few quarter-sawn boards. Plain-sawn logs are partly cut through and through; they produce a mixture of plain-sawn and rift-sawn boards.

Cutting quarter-sawn boards

There are a number of ways to cut a log so that it produces quarter-sawn boards. The ideal method is to cut each board parallel with the rays, like the radiating spokes of a wheel, but this is wasteful of timber and is not used commercially. The conventional method is to cut the log into quarters and convert each quadrant into boards. Commercial quarter-sawing first cuts the log into thick sections and then into quartered boards.

The end grain should be checked when selecting quarter-sawn wood; only those boards with growth rings at around 90 degrees to the surface should be chosen, to give the most stable wood. Not all timber merchants allow a buyer to select random boards; if they do, they will probably charge more.

STABILITY

Because wood shrinks as it dries, the shape of a board can change or 'move' as shrinkage takes place. In general, wood shrinks roughly twice as much along the line of the annual rings as it does across them. Tangentially cut plain-sawn boards shrink more in their width, and quarter-sawn boards shrink only slightly in width and very little in thickness.

Shrinkage can also cause boards to distort. The concentric growth rings of a tangentially cut plain-sawn board run approximately edge-to-edge and have different lengths; the longer outer rings shrink more than the inner rings, so that the board tends to bend or 'cup' across its width. Square sections of wood may become parallelogram-shape, and round sections can move to oval.

Because the growth rings of a quartered board run from face to face and are virtually the same length, less distortion takes place. This stability makes quartered boards the first choice for flooring and furniture-making.

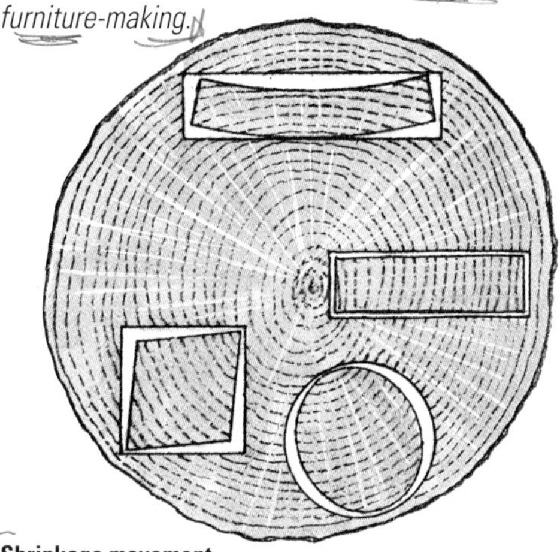

Shrinkage movement
Sections of wood distort differently, depending on growth-ring orientation.

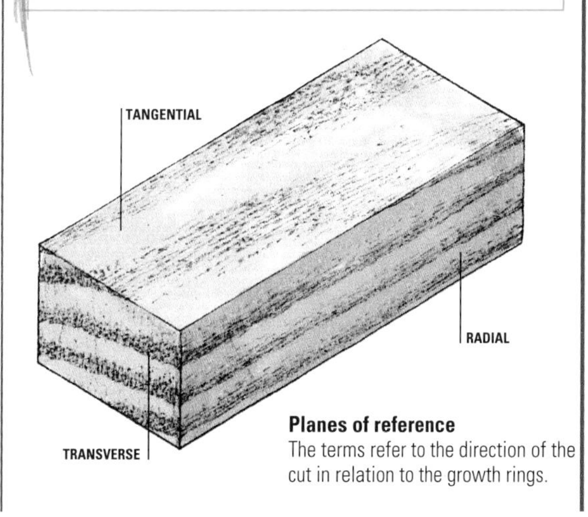

Planes of reference
The terms refer to the direction of the cut in relation to the growth rings.

MILLING YOUR OWN WOOD

NB

Logs can be converted into beams, boards or strips, or cut into 'blanks' or pieces for sculptors, turners or carvers. Commercial-timber merchants sell most sizes and sections, and specialist traders supply odd sizes and unusual woods, which can be relatively expensive. The development of portable sawmills means that dedicated woodworkers can convert their own wood.

Portable sawmills

Proprietary portable sawmills are powered by a chainsaw or bandsaw. A frame spans the width of a log, carrying a blade driven by a motor at one or both ends. Most machines use petrol-powered chainsaw engines, but electric-driven models can be used for lighter logs. The frame allows the blade to be raised or lowered to set the depth of cut.

The biggest advantage of a portable sawmill is that you can convert a log where it lies. Although these machines are relatively expensive, the money saved for new wood will soon cover the cost and can provide a useful sideline for supplying other woodworkers.

Types of sawmill

Chainsaw models are fitted with a ripping blade, which is made for cutting with the grain; the more powerful are capable of cutting large logs. This type should not be used for cutting across the grain.

The bandsaw blade cuts a thinner kerf than the chainsaw, and therefore produces less waste. One model can be used single-handed to produce material from 3 to 225mm (⅛ to 9in) thick, and has the capacity to cut a log 500mm (1ft 8in) wide.

Sources of wood

Trees for home milling can be found on land being cleared for development schemes, farmland or orchards, highways being maintained, or even local gardens and yards. These trees are unlikely to be highly valued, and can usually be bought for a low price; they may even be free. Whatever the source, only trees in good condition should be used. Logs from a site where nails or other metal fittings may have become embedded in them should be rejected, unless they are likely to produce some special wood.

Once the wood is cut, it needs to be seasoned before it is ready for use (see page 26).

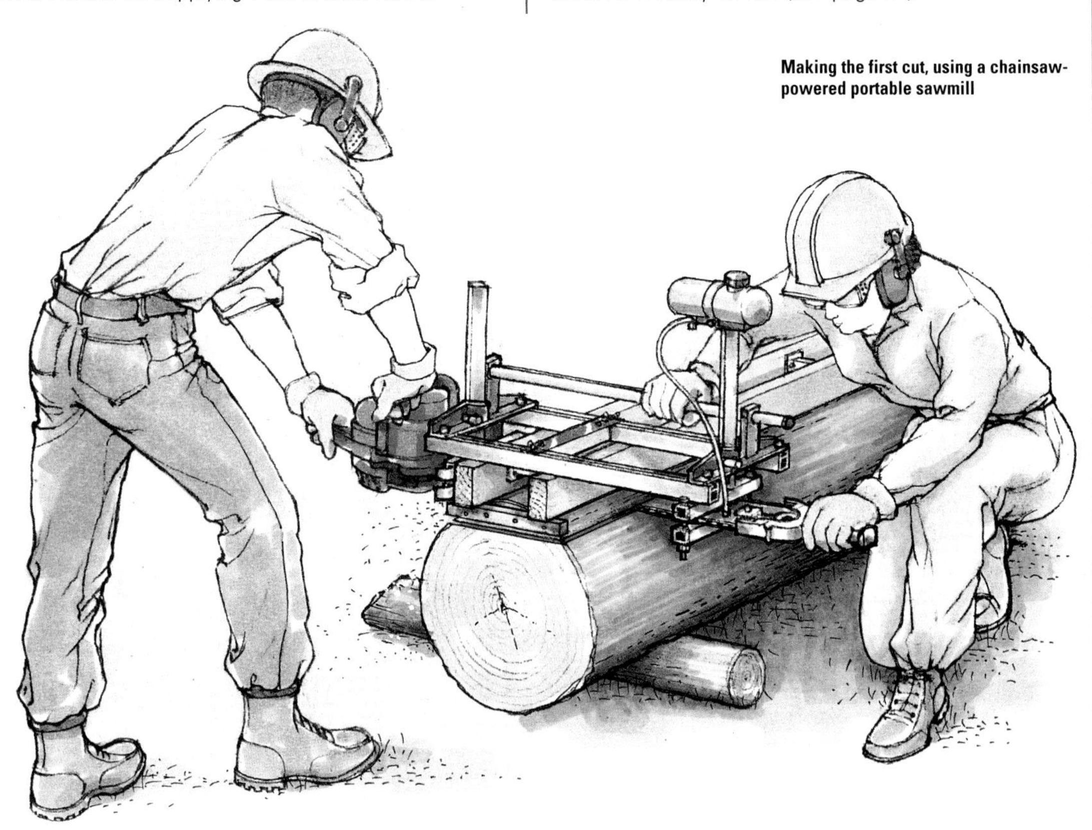

Making the first cut, using a chainsaw-powered portable sawmill

Preparing a log

The log should ideally be worked on level ground. All the limbs are first removed by trimming with a chainsaw. Care must be taken at all times, as trees are heavy and may lie awkwardly. The trimmed log can be cut to a suitable length, allowing for wastage, and the trimmings prepared for firewood.

Setting up

It is essential that the log is made immovable and stable. Large logs that are to be cut into thick sections can be worked on the ground once they are wedged firmly. It usually takes two people to manoeuvre heavy logs, using cant hooks. Trestles or V-blocks can also be used to support and steady lighter logs.

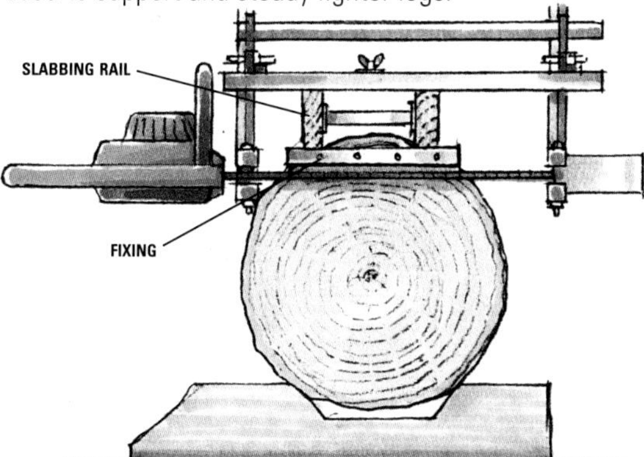

Making the first cut

A guide beam or 'slabbing rail', supplied with the sawmill or made as required, is fixed to the top of the log after any surface lumps have been trimmed away with a hatchet or saw, to allow the beam to lay flat. The depth of cut is set to clear the guide-beam fixings, and the sawmill is then slid along the beam to make the first, or 'slab', cut.

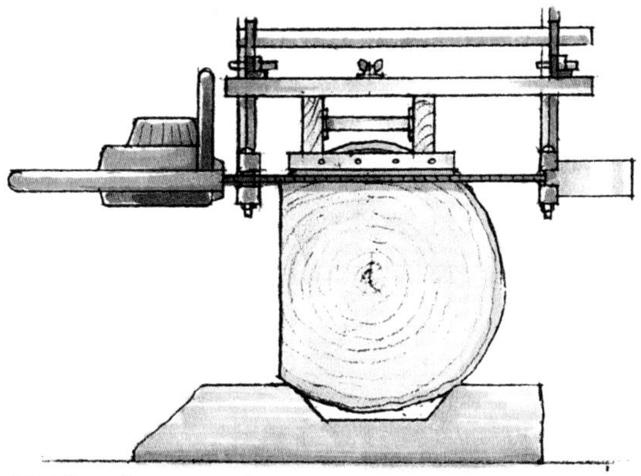

Making the second cut

The guide beam and cut slab are removed, and the log is turned through 90 degrees. The beam is then refitted and set at right angles to the first cut face of the log. The depth of cut may need to be re-set before the second cut is made in the same way as the first. The beam is then removed, along with the second slab.

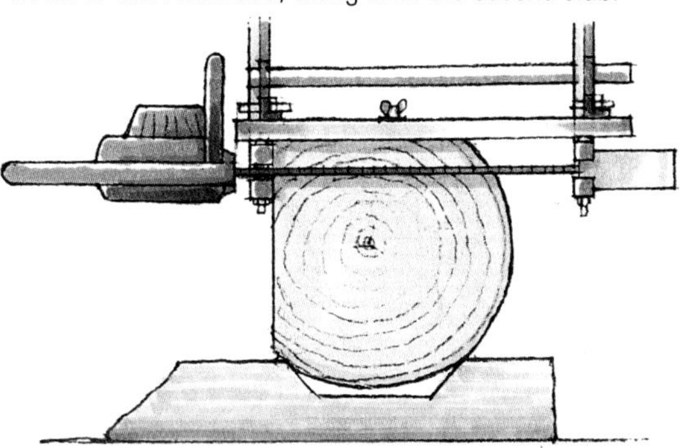

Cutting boards

The depth of cut is set to the required board thickness, and the cut is then made with the sawmill's guide rails sliding on the cut face of the wood. This process is used to produce boards with one square edge, and is repeated as each board is removed.

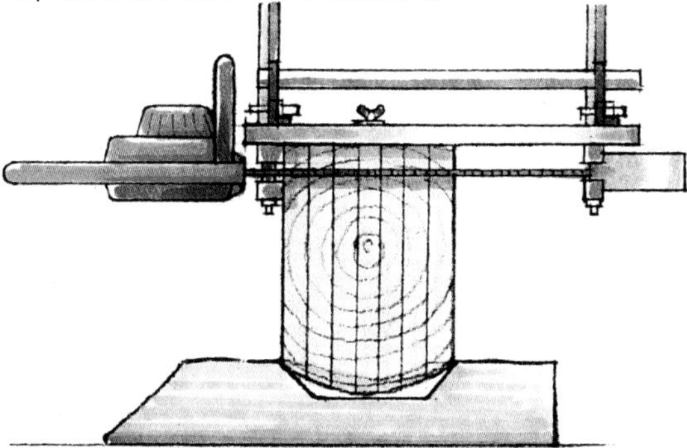

Cutting dimension stock

The log is first slab-cut on three sides and converted into boards. The boards are then placed on edge and clamped together. The saw is set, and the wood is sawn to the required width or thickness.

DRYING WOOD

Drying or 'seasoning' newly felled, or green, wood involves removing the free water and much of the bound moisture from the cell walls, in order to stabilize the wood. This process changes the wood's properties, increasing its density, stiffness and strength. Some carvers, sculptors and chair-makers work green wood, as it saves time and allows large pieces to be used, but seasoned timber is used for cabinetmaking and joinery.

Removing water content

In newly felled wood, the cell walls are saturated and the cell cavities hold free water. As the wood dries, free water evaporates from the cavities, but moisture remains within the cell walls – the fibre-saturation point; this occurs at about 30 per cent moisture content by weight (though this varies, depending on the species).

Shrinkage begins when the moisture is lost from the cell walls. When the moisture content is in balance with the relative humidity of its surroundings – known as the equilibrium moisture content (EMC) – the wood will stop losing water.

Seasoning must be carried out properly, to avoid stresses and to ensure that the EMC is at the appropriate level to prevent uneven swelling and shrinkage.

Commercial air-drying

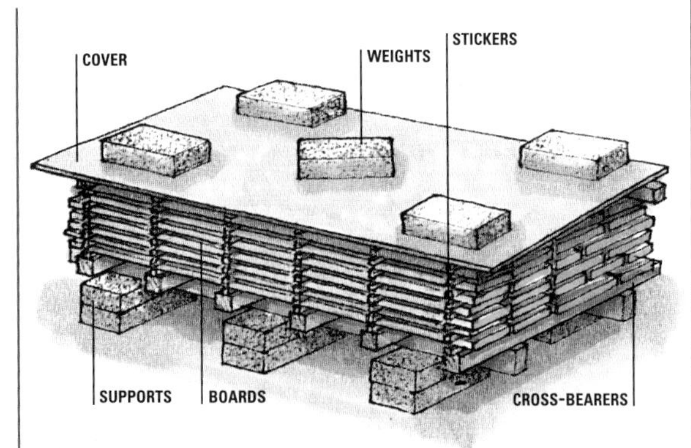

Stacking boards

Preparing wood for drying

Logs are best cut in winter, when low sap levels and temperature help to reduce fungal attack and shorten seasoning time. Boards cut through and through should be left with the bark and sapwood in place. This protects their edges from the elements and reduces distortion caused by rapid or uneven drying.

Air-drying

In this traditional method, stacks of wood are stored in ventilated sheds or in the open air, and the wood is dried by natural airflow through the stack. The boards are evenly stacked on 25mm (1in) square spacer battens called 'stickers', spaced 450mm (1ft 6in) apart. For hardwoods, it takes approximately one year to dry each 25mm (1in) thickness, and about half that time for drying softwoods.

Air-drying reduces the moisture content to about 14 to 16 per cent, depending on the ambient humidity. For interior use, the wood should then be dried in a kiln or restacked and left to dry naturally in the environment where it will be used.

Stacking boards

The site must provide good airflow and shelter from high winds, strong sunlight and heavy rain. (It is more important to protect the wood from sun than from rain.) The stack should be built clear of the ground on a level concrete base or on ground free from organic growth. Building blocks can be used to raise and support a base platform made from stout wooden bearers.

Cross-bearers are placed at the same spacing as the stickers; the boards are then positioned in even layers, and each sticker is aligned with the one below, to prevent distortion and bent boards in the stack. A sheet of waterproof plywood or a similar covering, which can be sloped to aid drainage, is laid over the top of the stack and weighted down. The ends of the boards are coated with a heavy sealing coat of paint, to prevent them splitting due to rapid drying.

Loading stacks into a large commercial kiln

Kiln-drying

Wood intended for interior use needs a moisture content of about 8 to 10 per cent, sometimes lower. The particular advantage of kiln-drying is that it takes only days or weeks to reduce the moisture content of wood below air-dry levels; however, some woodworkers prefer to work air-dried wood. Kiln-drying can change the colour of some woods – for example, beech takes on a pink shade when kiln-dried.

Stickered stacks of boards are loaded onto trolleys and rolled into the kiln. A carefully controlled mixture of hot air and steam is pumped through the piled wood, and the humidity is gradually reduced to a specified moisture content. Wood dried below air-dry levels will try to take up moisture if left exposed, so kiln-dried wood is best stored in the environment in which it is to be used.

Preparing a compact home kiln

Kiln-drying at home

Woodworkers who wish to control all aspects of their work can purchase kilns for drying their own wood. The smallest 'home' unit is 1.2 x 1.2 x 2.7m (4 x 4 x 9ft) and uses a dry heat-vent system run from a domestic power supply. Electronic temperature and humidity controls can be adjusted on a daily basis, based on the manufacturer's schedules. Fan-assisted electric heaters raise the temperature and blow hot air around a stickered stack of boards built on a loading cart. Moisture from the wood is circulated within the chamber and vented at a controlled rate, gradually drying the wood without degrading it.

Checking moisture content

The moisture content of wood is given as a percentage of its oven-dry weight. This is calculated by comparing the original weight of a sample block of newly felled wood (preferably taken from the middle of the board rather than the ends, which may be drier) with the weight of the same sample after it has been fully dried in an oven. To find the lost weight, subtract the dry weight from the original weight. The following formula is used to calculate the moisture-content percentage:

$$\frac{\text{Weight of water lost from sample}}{\text{Oven-dry weight of sample}} \times 100$$

Moisture meters

A moisture meter is a simple and convenient tool for checking moisture content. The meter measures the electrical resistance of the moist wood and provides an instant reading of the moisture-content percentage. Standard moisture meters have two-pin electrodes, which are inserted into the wood, while pinless meters use an electromagnetic system to give a moisture reading to a depth of 18mm (¾in) by holding the meter over the surface of the board. Pinless meters do not mark the wood, and can be used to check finished parts.

When using either type of meter, it is important to check the moisture level at various points along the wood, as not all parts of a board dry at the same rate.

SELECTING WOOD

The selection of a suitable wood for a project is usually based on the appearance of the material and its physical and working properties. When the species has been selected, the boards, ideally from the same tree, are then chosen for quality and condition. Lastly, the wood is evaluated at the making stage, in order to explore its full potential in the finished work.

Buying wood

Timber suppliers usually stock the softwoods most commonly used for carpentry and joinery – spruce, fir and pine. They are generally sold as 'dimension' or 'dressed' stock, the trade terms for sawn or surface-planed sections cut to standard sizes. One or more of the faces may be surfaced.

Most hardwoods are sold as boards of random width and length, although some species can be bought as dimension stock. Dimension timber is sold in 1ft or 300mm units (see right); check which system your supplier uses, as the metric unit is about 5mm (3/16in) shorter than an imperial foot. Whichever system you use, always allow extra length for waste and selection.

When working out timber requirements, remember that planing processes can remove at least 3mm (1/8in) from each face of the wood, making the actual width and thickness less than the 'nominal' or 'sawn size' quoted by the timber merchant. The length, however, is always as quoted.

Stacks of boards at a timber yard

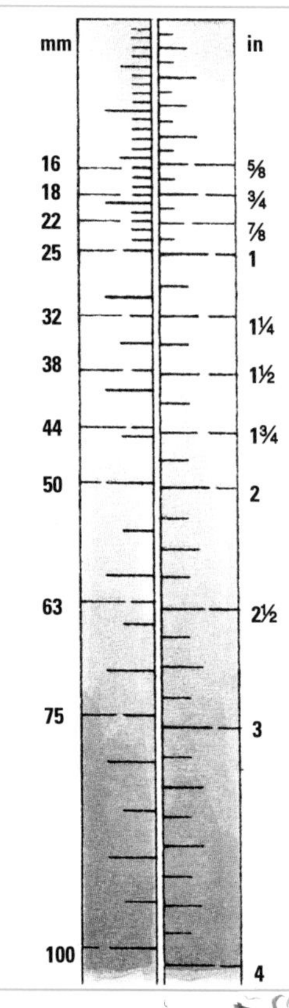

IMPERIAL AND METRIC MEASUREMENTS

The timber trade is an international business, in which the producer countries use either the imperial or metric systems of measurement. Though there are moves for unification towards the metric system, both are in use at time of writing. To avoid confusion and inaccuracies, only one system should be used when specifying dimensions. This actual-size bar chart shows the slight variation for standard sizes up to 100mm (4in).

Grading woods

Softwoods are graded for evenness of grain and the amount of allowable defects, such as knots. For general woodworking, better-quality 'appearance' and 'non-stress' grades are probably the most useful. Stress-graded softwoods are rated for structural use where strength is important. The trade term 'clear timber' is used for knot-free or defect-free wood, but is not usually available from suppliers unless specified.

Hardwoods are graded by the area of defect-free wood: the greater the area, the higher the grade. The most suitable grades for general woodworking are 'firsts' and 'FAS' – 'firsts and seconds'.

Although specialist firms supply wood by mail order, personal selection is by far the best option. Take a block plane with you when buying timber, so you can expose a small area of wood if the colour and grain are obscured by dirt or saw marks.

Cutting lists

Cutting lists are used to specify the finished length, width and thickness of every component of a project. The list should also state the material and quantity required. A cutting list enables a timber merchant to supply the material in the most economic way, and provides the woodworker with a schedule for converting the wood to size.

Defects in wood

If wood is not dried carefully, stresses can mar it or make it difficult to work. Insufficient drying can cause shrinkage of dimensioned parts, joints opening, warping and splitting. Check the surface for obvious faults, such as splits, knots and uneven grain. Look at the end section, to identify how the wood was cut from the log, and to spot any distortion. Sight along the length to test for twisting or bowing. Ingrained stains, caused by water collecting in the stack or the use of incompatible wood for the stickers, can be difficult to remove, so check for sticker marks. In addition, look for evidence of insect attack or traces of fungal growth.

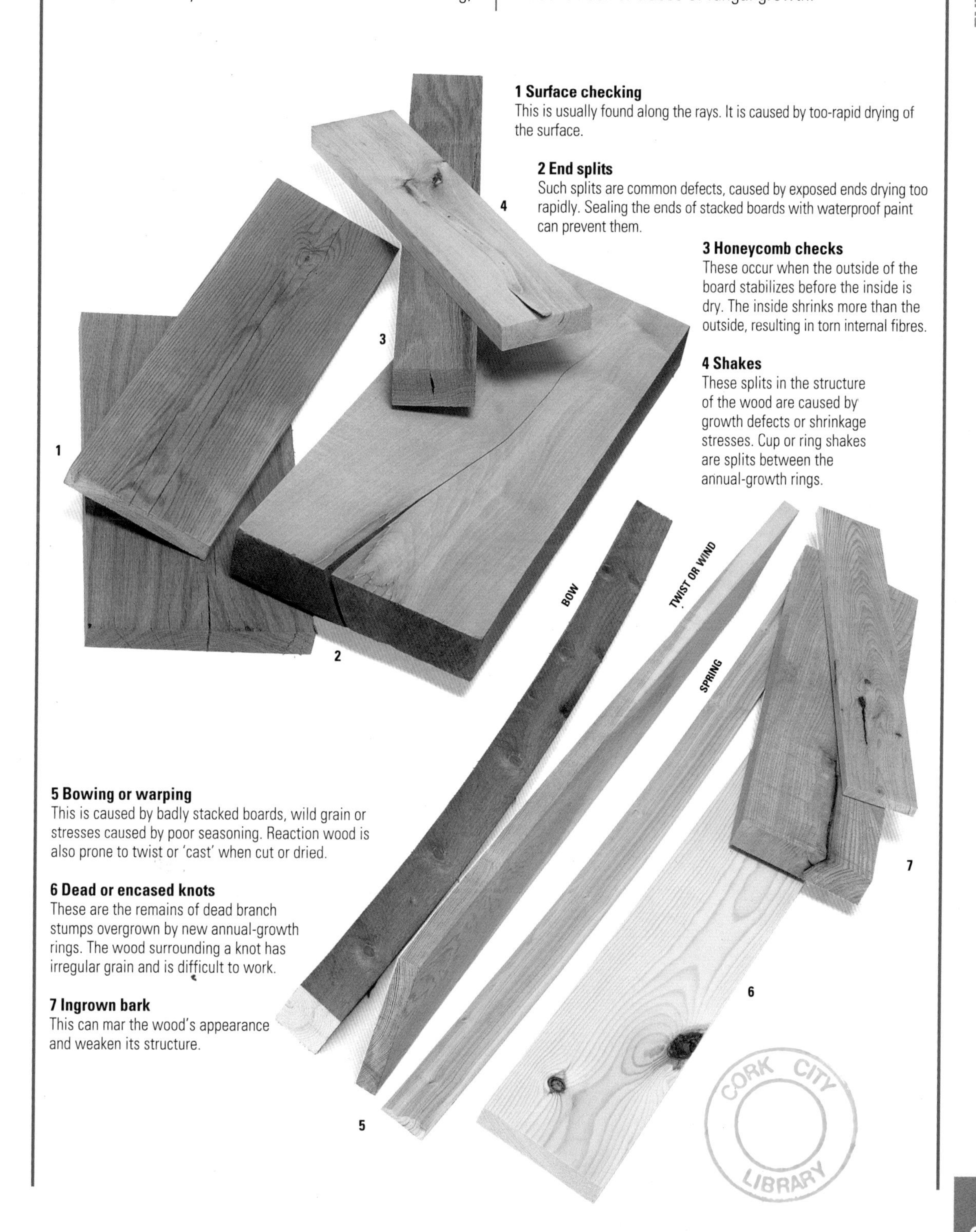

1 Surface checking
This is usually found along the rays. It is caused by too-rapid drying of the surface.

2 End splits
Such splits are common defects, caused by exposed ends drying too rapidly. Sealing the ends of stacked boards with waterproof paint can prevent them.

3 Honeycomb checks
These occur when the outside of the board stabilizes before the inside is dry. The inside shrinks more than the outside, resulting in torn internal fibres.

4 Shakes
These splits in the structure of the wood are caused by growth defects or shrinkage stresses. Cup or ring shakes are splits between the annual-growth rings.

5 Bowing or warping
This is caused by badly stacked boards, wild grain or stresses caused by poor seasoning. Reaction wood is also prone to twist or 'cast' when cut or dried.

6 Dead or encased knots
These are the remains of dead branch stumps overgrown by new annual-growth rings. The wood surrounding a knot has irregular grain and is difficult to work.

7 Ingrown bark
This can mar the wood's appearance and weaken its structure.

PROPERTIES OF WOOD

In many woodworking projects, the grain pattern, colour and texture are the most important factors when choosing which woods to work. Though equally important, the strength and working characteristics are often a secondary consideration – and when using veneer, the appearance is all.

Working wood is a constant process of discovery and learning. Each piece is unique, and any section of wood taken from the same tree, even from the same board, will be different and a challenge to the woodworker's skills. You can only gain a full understanding of wood's properties by working it and experiencing the way it behaves.

Grain

The mass of the wood's cell structure constitutes the grain of the wood, which follows the main axis of the tree's trunk. The disposition and degree of orientation of the longitudinal cells create different types of grain.

Trees that grow straight and even produce straight-grained wood. When cells deviate from the main axis of the tree, they produce cross-grained wood. Spiral grain comes from trees that twist as they grow; when this spiral growth changes direction from one angle to another, each change taking place over a few annual-growth rings, the result is interlocked grain. Wavy grain, which has short, even waves, and irregular curly grain occur in trees with an undulating cell structure. Wild grain is created when the cells change direction throughout the wood; irregular-grained woods of this kind can be difficult to work.

Random and undulating grain make various patterns in wood, according to the angle to the surface and light reflectivity of the cell structure. Boards with these configurations are particularly valued for veneer.

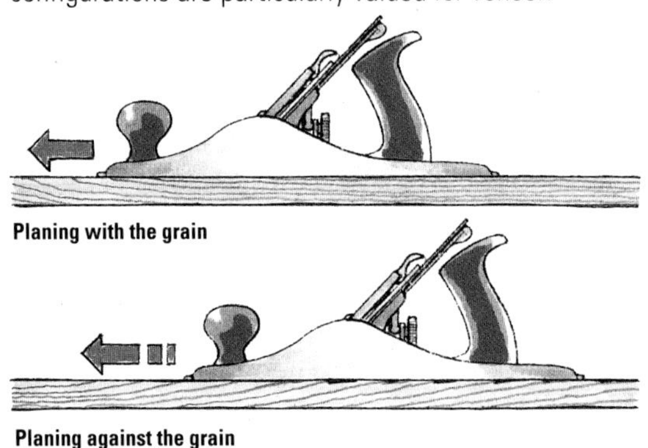

Planing with the grain

Planing against the grain

Working wood

Planing 'with the grain' follows the direction of the grain where the fibres are parallel or slope up and away from the direction of the cutting action, resulting in smooth, trouble-free cuts. Planing a surface 'against the grain' refers to cuts made where the fibres slope up and towards the direction of the planing action; this produces a rough cut. Sawing 'with the grain' means cutting along the length of the wood, in the same direction as the longitudinal cells. Sawing or planing 'across the grain' describes cuts made more or less perpendicular to the grain of the wood.

Figure

The term 'grain' is also used to describe the appearance of wood; however, what is really being referred to is a combination of natural features collectively known as the figure. These features include the difference in growth between the earlywood and latewood, the way colour is distributed, the density, concentricity or eccentricity of the annual-growth rings, the effect of disease or damage, and how the wood is converted.

Using figure

When tree trunks are tangentially cut, the plain-sawn boards display a U-shape pattern. When the trunk is radially cut or quarter-sawn, the series of parallel lines usually produces a less distinctive pattern.

The fork where the main stem of the tree and a branch meet provides curl, or crotch, figure that is sought-after for veneer. Burr wood, an abnormal growth on the side of a tree caused by injury, is used for veneer. It is popular among woodturners, as is the random-grain figure of stumpwood, from the base of the trunk or roots.

Texture

Texture refers to the relative size of the wood's cells. Fine-textured woods have small, closely spaced cells, while coarse-textured woods have relatively large cells. Texture also denotes the distribution of the cells in relation to the annual-growth rings. A wood where the difference between earlywood and latewood is marked has an uneven texture; one with only slight contrast in the growth rings is even-textured.

Coarse-textured woods, such as oak or ash, tend to have finer cells when they are slow-grown, and are also lighter and softer than when fast-grown. Fast-grown trees usually produce a more distinctive figure and harder, stronger and heavier wood.

Effects of texture

The difference in texture between earlywood and latewood is important to the woodworker, as lighter-weight earlywood is easier to cut than the denser latewood. If tool-cutting edges are kept sharp, this should minimize any problems, but latewood can be left proud of the earlywood when finished with a power sander. Those woods with even-textured growth rings are generally the easiest to work and finish.

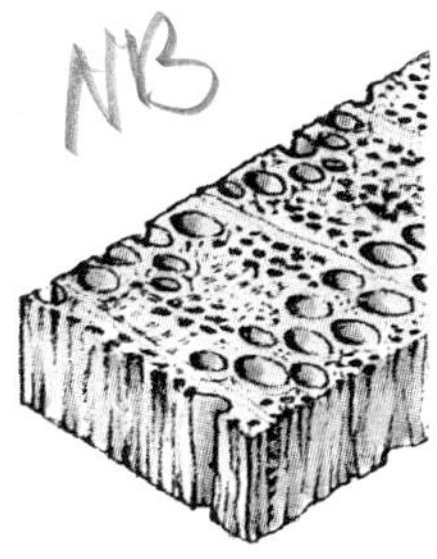

Ring-porous wood

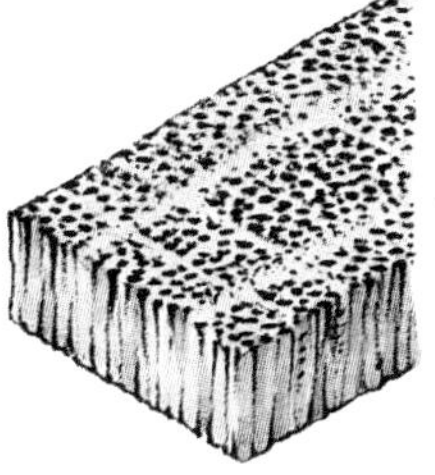

Diffuse-porous wood

Hardwood porosity

The distribution of hardwood cells can have a marked effect on wood texture. The 'ring-porous' hardwoods, such as oak or ash, have clearly defined rings of large vessels in the earlywood, and dense fibres and cell tissue in the latewood; this makes them more difficult to finish than the 'diffuse-porous' woods, such as beech, where the vessels and fibres are relatively evenly distributed. Woods like mahogany can be diffuse-porous, but their larger cells can make them more coarse-textured.

Durability

Durability refers to a wood's performance when it is in contact with soil. Perishable wood is rated at less than 5 years, very durable at more than 25 years. The durability of a species can vary according to the level of exposure and climatic conditions.

BOTANICAL CLASSIFICATION

In the sections on softwoods and hardwoods of the world (see pages 44–53 and 56–82), each wood is listed in alphabetical order according to the botanical classification of the genus and species, which is shown in italics beneath the most commonly used commercial name.

The botanical name is the only universal name which can be relied on to accurately identify a species of wood. In suppliers' catalogues and reference books, as here, the terms 'sp.' or 'spp.' are commonly used to indicate that a wood may be one of a variety of species within a genus, or 'family', of trees.

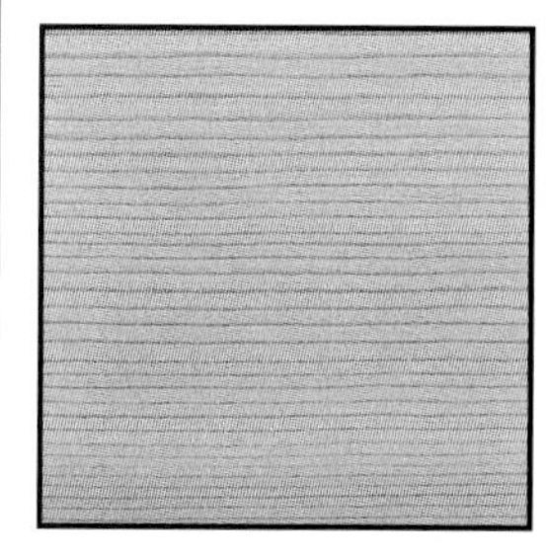

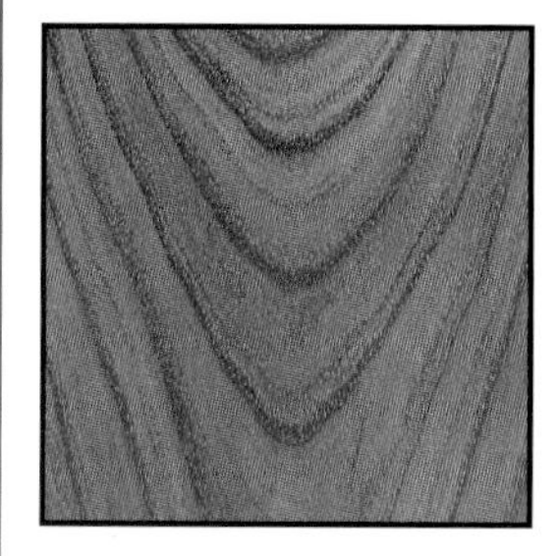
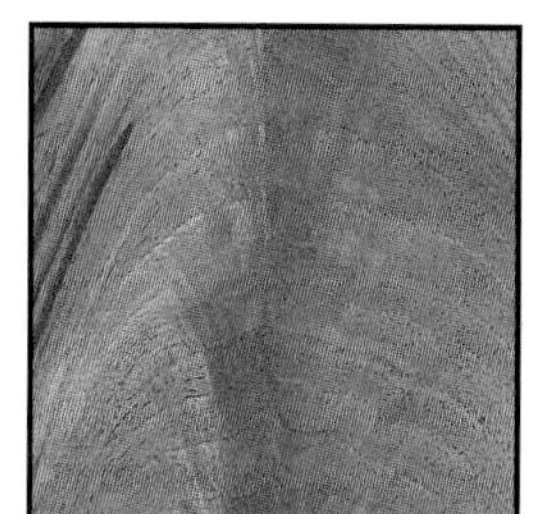
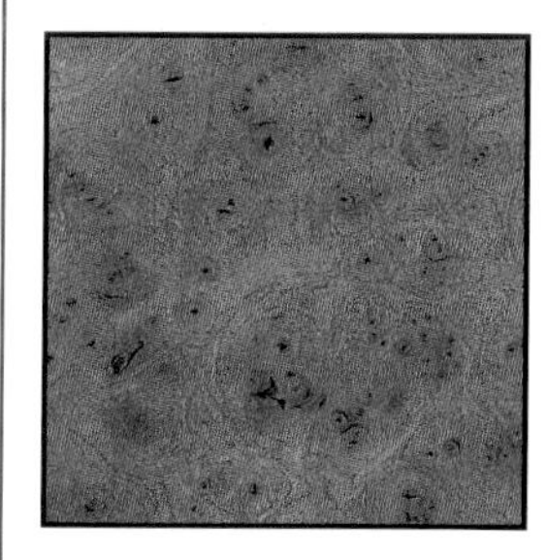
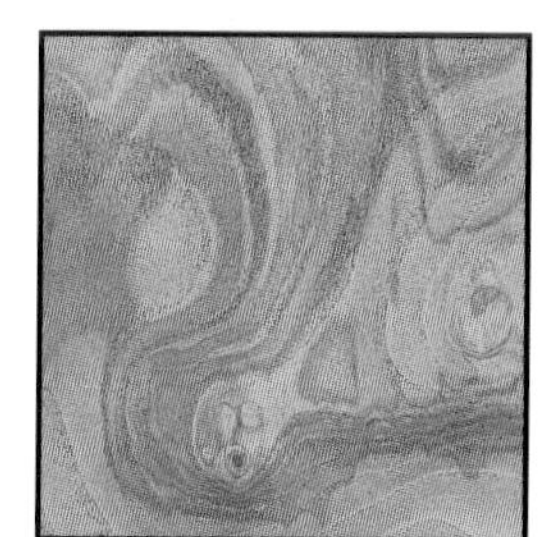

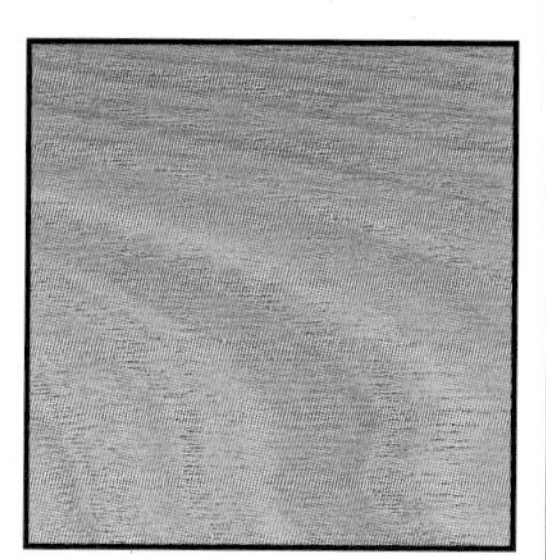

Textures and patterns (from left to right, in sequence):
Straight grain (sitka spruce)
Spiral grain (satinwood)
Wavy grain (fiddleback sycamore)
Wild grain (yellow birch)
U-shape pattern (blackwood)
Curl or crotch (walnut)
Burr wood (elm)
Stumpwood (ash)
Fine-textured (lime)
Coarse-textured (sweet chestnut)

STEAM-BENDING WOOD

Wood's versatility is rarely better shown than when it is bent into curves, to provide both structural strength and decoration. Thinly cut wood can be bent into gentle curves without being treated beforehand; tighter bends and thicker wood require the use of a basic steam chest, shaped former and support strap. Steam softens the wood fibres, to allow them to bend and compress around the former, and the strap restrains the outer fibres and prevents them from splitting out.

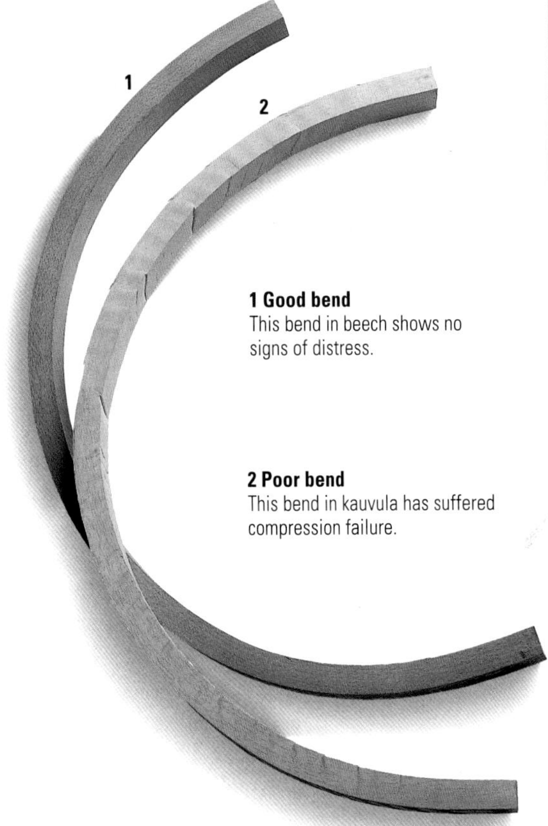

1 Good bend
This bend in beech shows no signs of distress.

2 Poor bend
This bend in kauvula has suffered compression failure.

Choosing wood

Because any flaw in wood is a potential weakness when bending, only straight-grained timber that is free from knots and shakes should be used. Newly cut green timber is easiest to bend, and air-dried seasoned wood is easier than kiln-dried; it is preferable to soak dry and difficult wood for a few hours before steaming it.

The lists of woods on pages 44–53 and 56–82 indicate those that are particularly good for steam-bending, and there are many others. However, bending is not an exact science, even with flaw-free wood, and a few failures can be expected along the way.

Preparing wood

Whether prepared before or after bending, and whatever the size or shape of the section, the wood must always be cut about 100mm (4in) over-length before bending; any end splits or compression damage from the strap can be cut away after bending.

Wood that has been brought to a smooth finish is less likely to split after being bent, and makes final finishing an easier process. Prepared green wood shrinks more than seasoned timber; green wood turned to a round section before bending can dry oval in shape.

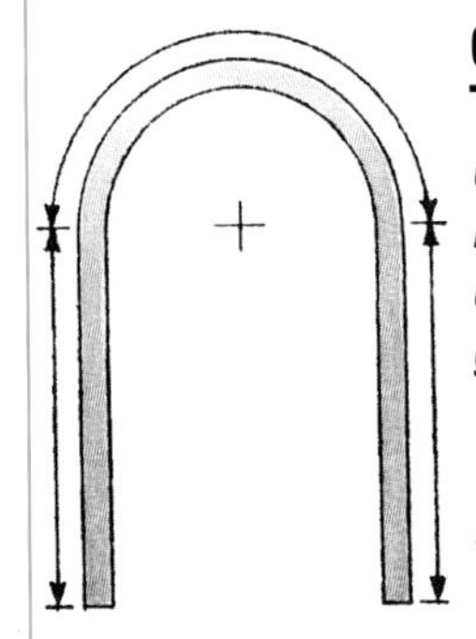

CALCULATING THE LENGTH

On a full-size drawing of the required final shape, the outside edge is measured to give the length for the piece; the inner fibres of the wood will compress and take up the smaller inner curve.

The strap

The strap should be made from 1.5mm (1⁄16in) thick flexible mild steel, wider than the wood to be bent; plated or stainless steel prevents chemical stains, or the strap can be wrapped in polythene, which also aids insulation. Hardwood or metal end stops restrain the ends of the wood and prevent it from stretching or splitting on the outside of the curve. They must be strong, and large enough to support the end of the wood.

The distance between the stops should equal the length of the work, including the waste, and the strap is marked and drilled to take end-stop bolts. Hardwood levers, attached to the back face of the strap with the end-stop bolts, provide extra leverage for bending the strap and the steamed wood around the former.

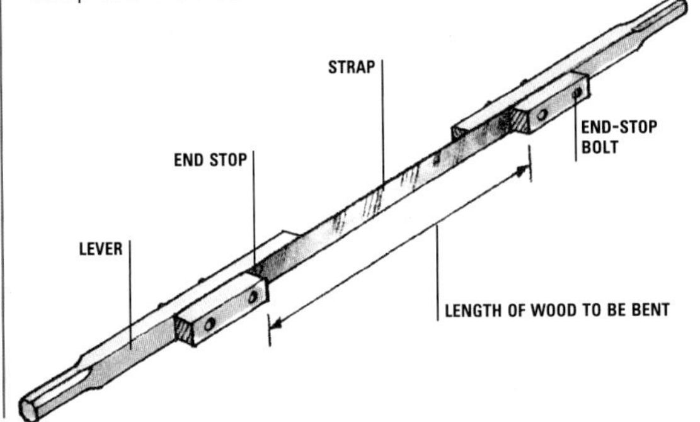

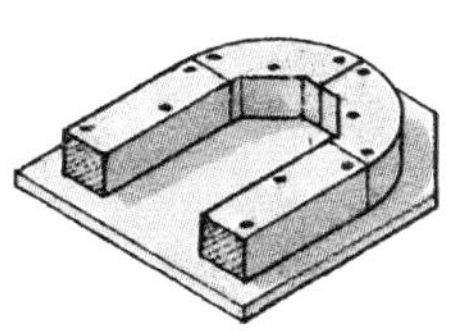
Solid-wood former

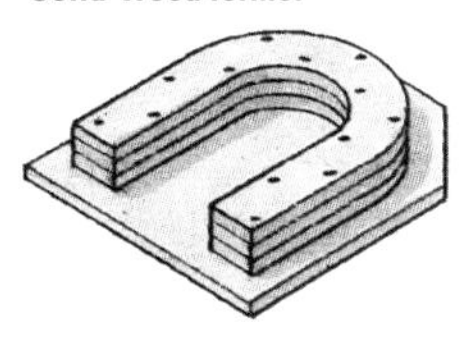
Plywood former

Making a former

The former sets the final shape of the steamed wood and supports the softened inner fibres. It should be at least as wide as the wood to be bent, and must incorporate locations for the cramps that hold the strap and work. The shape of the former must allow for bent wood's tendency to straighten when released from clamping pressure; this is usually worked out by experiment. Thick sections of wood, layers of glued-together plywood, or a shaped plywood panel and frame, can be mounted on a man-made-board base.

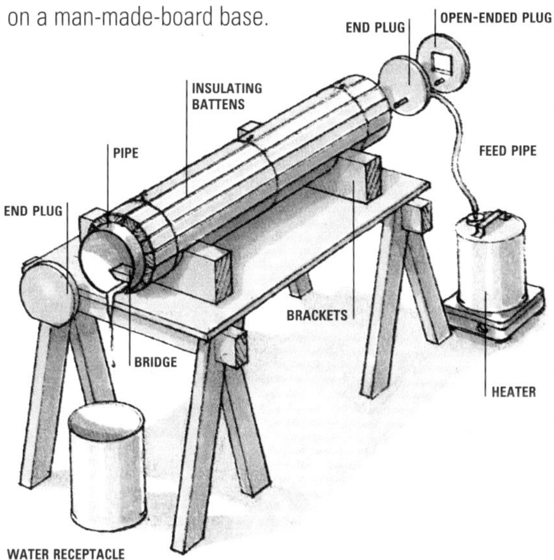

Making a steam chest

A length of metal or plastic pipe can be used to make a chest that holds small pieces of wood, or takes longer pieces that only require bending along part of their length. Removable end plugs, made from exterior-grade plywood, are push-fitted to either end; one has a hole for the steam-feed pipe, and a vent and a drainage hole are planed flat in the bottom edge of the other. Open-ended plugs allow longer pieces to be passed through, and wooden bridges fitted inside the pipe keep shorter pieces of wood off the bottom.

To insulate the pipe, plastic foam or wooden battens are held in position by metal ties. Support brackets hold the pipe at a slight angle, to allow condensation to drain out as water into a receptacle.

Alternatively, a simple glued-and-screwed box-type steam chest can be constructed from exterior-grade plywood. This type of chest can be built to exact sizes, and is the best option for steaming batches of wood.

Building up steam

You can use a small electric-powered boiler to generate steam, or make your own from a medium-size metal drum fitted with a removable cap or plug. This can be heated by a portable gas burner or an electric hot plate. One end of a short hose is fitted to a spigot soldered into the drum, with the other end plugged into the end plug of the steam chest. A continuous supply of steam is produced by half-filling the drum and heating the water to 100°C (212°F).

As a guide, the wood should be steamed for one hour for every 25mm (1in) of thickness. The wood must not be left in for too long, as this can break down the structure of the wood, and will not necessarily improve the bending qualities.

Preparation for bending

Because steamed wood can only be worked into a bent shape for a few minutes before cooling and setting, make sure that everything is ready before starting. The metal strap must be warmed so that it doesn't cool the wood on contact, and it is vital that there are enough cramps to hand. When bending thick material, it is a good idea to ask a friend to help.

Bending the wood

With the steam generator shut down, the wood is removed from the steam chest and placed in the pre-warmed strap. The assembly is set on the former and clamped at the centre, with a block of waste wood between the cramp and strap. Once the wood is bent into shape around the former, it is clamped securely in position, again using waste-wood blocks where necessary. The wood can be left to cool and set on the former, or removed after at least 15 minutes and re-clamped to a drying jig. It will need to dry for up to a week.

Clamping on the former

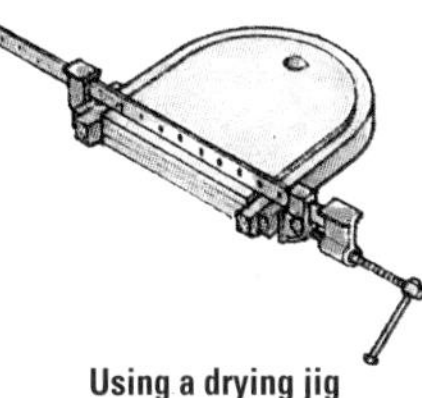
Using a drying jig

BENDING LAMINATED WOOD

Unlike solid wood, which needs steam and heat to be bent, the thin strips of wood or veneer used for laminating can be bent dry round one or more formers and glued together to make a solid shape. Because the fibres of each strip are not over-stressed, laminated components can be bent into tighter or more complex curves than can be produced from steamed solid wood.

Laminated-beech chair

Choosing wood

Strips of commercially produced 'construction' veneer can be laminated for making frame members, such as table or chair legs, and random-grain decorative veneer can be used for face laminates. When making your own from solid wood, select wood that is straight-grained and free from knots or shakes. As with steam-bending, air-dried wood is better than kiln-dried timber, because it is less brittle and bends more readily.

Once it is cut sufficiently thin, almost any wood can be used in laminates. The woods recommended for steam-bending on pages 44–53 and 56–82 are among those with the best pliability.

Cutting wood strips

For a consistent grain pattern which provides optimum reliability, the best option is cutting laminate strips from solid wood. Quarter-sawn boards are best for the purpose, as the growth rings run across the width of the strips, making them easier to bend. Before the board is cut, marking the face or end with a V-shape reference line helps realign the strips when gluing.

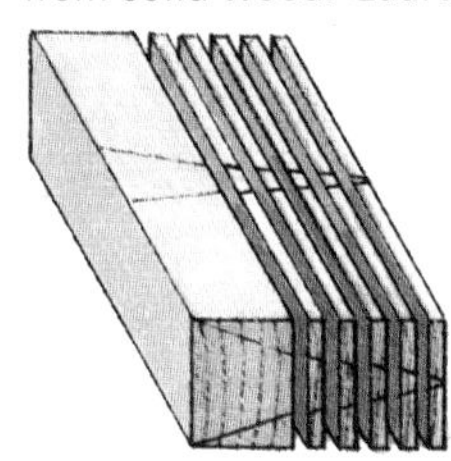

THICK AND THIN STRIPS

Thin strips of veneer or solid wood can be bent tightly and are unlikely to spring back from a bend. When cutting strips, the kerf of the saw blade wastes wood with each cut, so it is more economical to cut the strips as thick as can be worked. Different thicknesses of strips can be tested to see how well each one bends.

When using relatively thick strips, or when making tight bends, the wood can be bent before gluing by being dampened and placed in the appropriate former until it is dry.

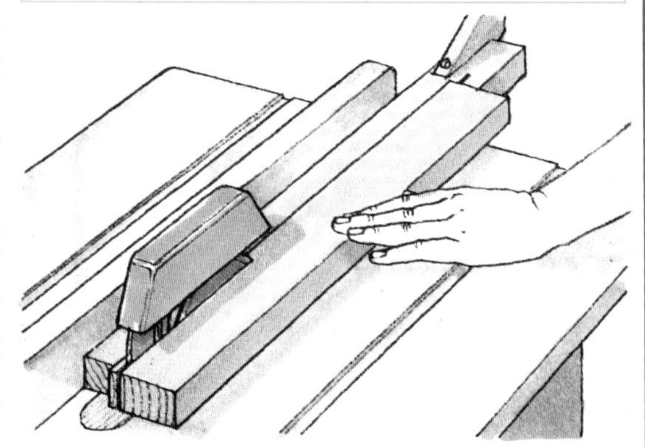

Using power saws

When cutting wood strips on a bandsaw, the planed edge of the board is passed against the fence, cutting a slightly thicker strip than is needed. The new cut edge of the board is planed, and the process repeated. The strips are finished by running through a thicknesser.

A table saw can be used to produce finished strips, but the saw may snatch and break the cut pieces, or even throw them back. A guide batten fitted with an end stop pushes the board along the fence to cut very thin strips; there must not be a wide gap between the blade and the table insert.

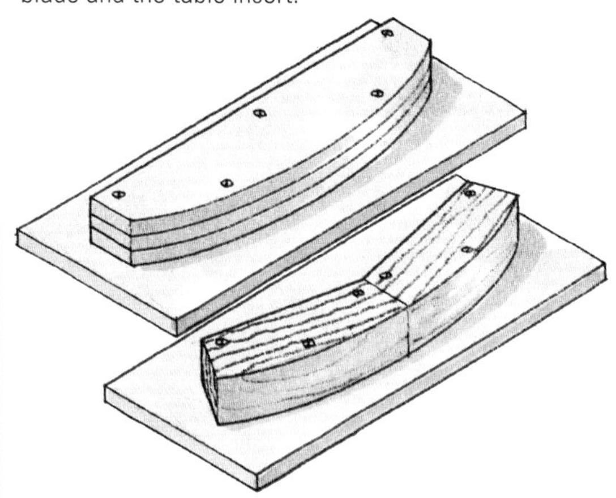

Particle-board former (top) and solid-wood former (bottom)

Formers
Glued laminated strips are clamped against single 'male' or matched 'male and female' formers until set.

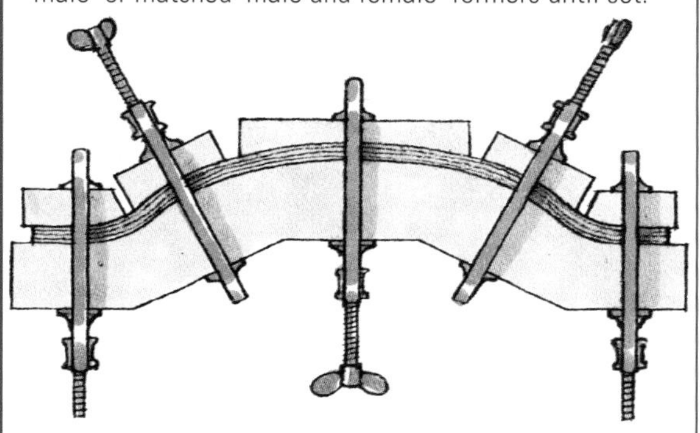

Making single formers
The contour is marked onto the face of a block of solid wood or layered particle board, and is cut on a band-saw; the profiled face must be longer than the wood strips. The cramps should ideally be positioned at right angles to the former face, so the back of the former should approximate the face contour.

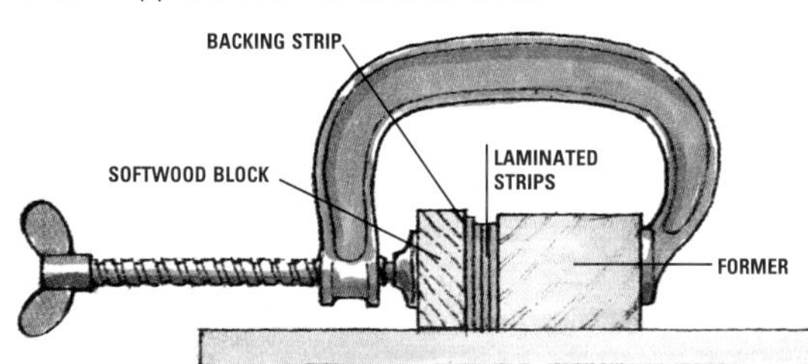

Protecting the laminate
Indentations in the top laminate can be avoided by covering it with a strip of spare laminate or waxed hardboard, and by fitting softwood blocks under the cramp heads, to evenly distribute the clamping load.

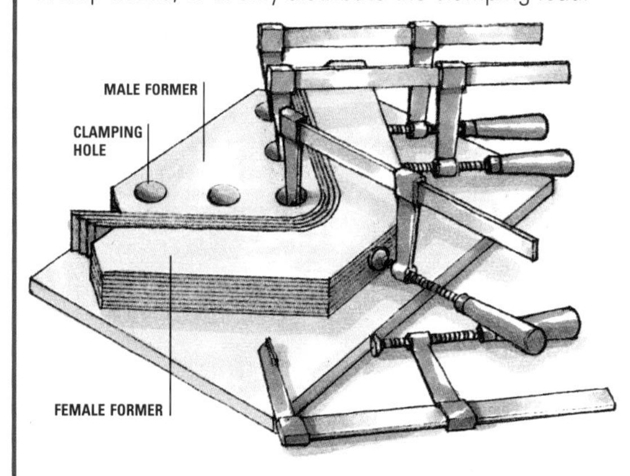

Making two-part formers
Matched male and female formers create even pressure across the area of the moulding. The formers are held together by a number of cramps inserted in holes drilled in the male former. Both formers can be made from solid wood or layered man-made boards.

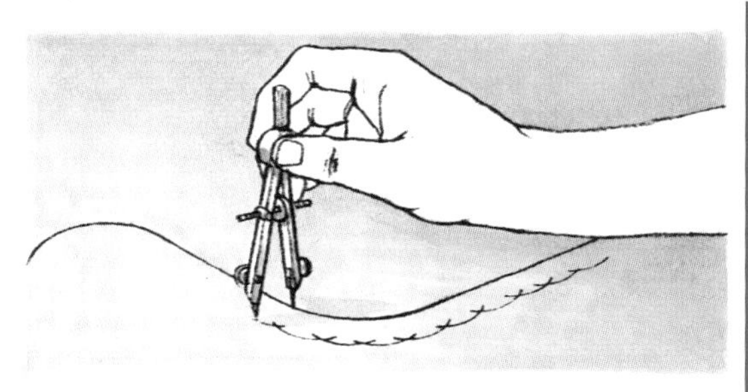

Marking the contour lines
The contour faces of male and female formers are produced by cutting along two parallel lines; using a single cut to form both contours will not work. The exact width between the parallel lines is determined by clamping together the wood or veneer strips to be glued, and measuring the thickness. The external and internal radii of bends that follow the curves of a compass are marked and then cut out.

A little more work is required for freehand or random bends. With one contour line marked out on the former, a pair of compasses makes a series of closely spaced arcs equal to the thickness of the laminate. The second contour line touches the peak of each arc.

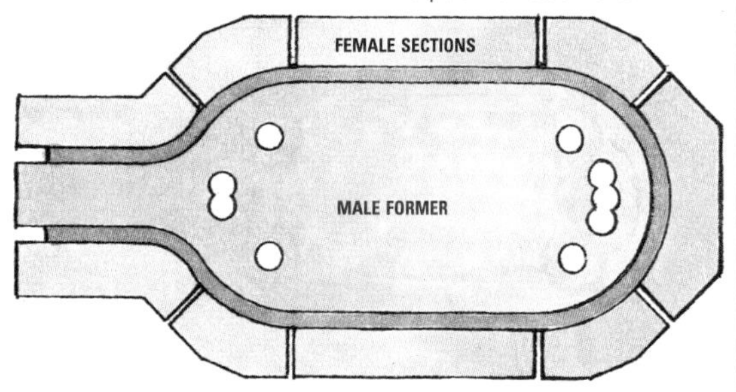

Multiple-part formers
Single or two-part formers are all that are required for most laminate-bending work. However, when the male former needs to be undercut to make a shape, the female former is cut into multiple sections, making assembly and removal of the laminated parts easier.

GLUING
The best adhesive for laminating is urea-formaldehyde glue, which sets slowly, allowing time to assemble the strips in position, and is less likely to creep than PVA glues. An even coating of glue is brushed onto the faces of each strip; the strips are then re-stacked in reverse order. The stacks are positioned on a single former or in two-part formers and clamped together, starting from the centre and working outwards, with even pressure applied to extrude glue and air. Bands of self-adhesive tape keep wide laminates together in male and female formers.

THE VERSATILITY OF WOOD

The uses to which wood can be put seem endless. So common has it become in our everyday environment that it is often taken for granted and hardly recognized for its value. The diversity of woods, with their variety of properties, ensures a wide selection for the woodworker. However, the wonder of wood is not just in its availability or variety; its workability with even the simplest of handtools has elevated it to the most used raw material of all.

With the development of edged tools, the human species has been able to fashion wood to change and enhance its environment – one only has to look at the history of all cultures to see examples of wooden artefacts and structures. Even with the development of synthetic materials and the progress of automated, mechanized production of wood and wood products, the raw material is still processed by traditional methods to meet a never-ending demand for products made from this most desirable natural material.

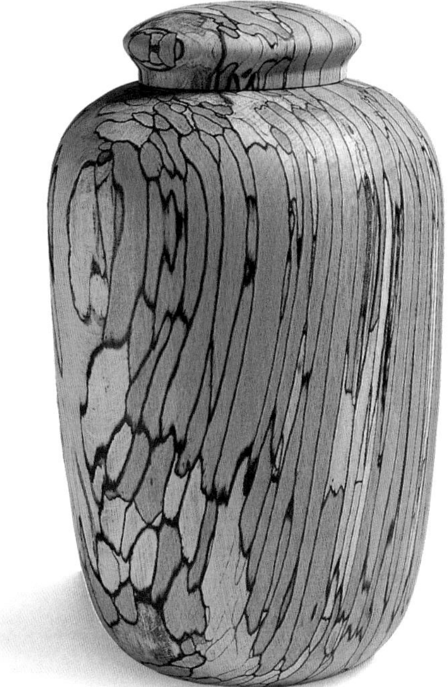

Lidded container
Spalted wood, which is caused by fungal attack, is much prized by woodturners for its incredible decorative patterns. In this example, the black 'zone lines' and mottled colouring that penetrate the wood produce a unique random design that is exploited by the woodturner.

Ship frames
Oak has long been used in traditional building construction and ship-building. Here, massive curved-oak frames are fitted to a keel to construct a replica of John Cabot's ship, the *Matthew*, the original crossed the Atlantic in 1497.

Windsor chair
Typically made of turned spindles, steam-bent bows and solid, shaped seats, traditional Windsor chairs are classic examples of the chairmaker's art. They are made in various regional styles, using native woods – such as ash, elm, yew, oak, beech, birch, maple or poplar – and can be found, in original or reproduced forms, in homes around the world.

'Seal table'
The natural colour and texture of a 'found' piece of European-sycamore log are creatively transformed into this delightful carved seal. The wood also forms the base for a clear-glass table top that represents the surface of water.

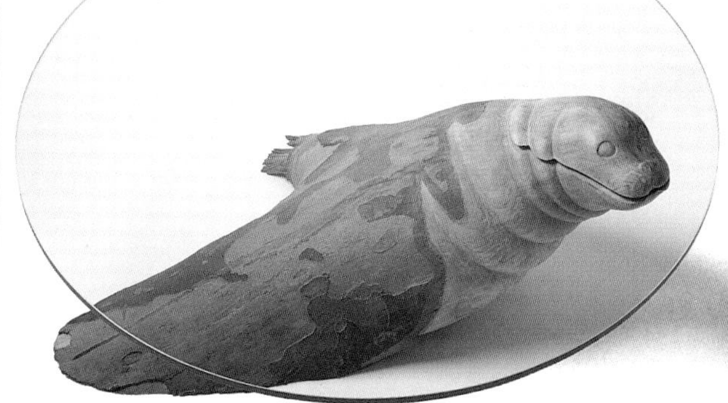

Burr bowl
Solid burr wood is a favourite material for woodturners. In this striking example, the natural contours and textures of elm burr are accentuated by flaming the workpiece with a blowtorch during turning; the turned grooves and smooth inner surface add textural contrast.

Shaker box
The simple design and fine craftsmanship of the American Shaker sect are clearly seen in this handmade traditional oval box. Thin-cut cherry wood is steamed and bent round a former before the projecting 'fingers' are secured by copper rivets; solid-wood ovals are then pinned into the lid and body.

WOOD'S ENEMIES

Wood is not only a renewable resource, but is also biodegradable. Sunlight, water and biological agents combine to sustain the living tree; once the tree is cut down, these life-giving elements cause wood left in the forest to degrade and, combined with insect attack, break down the wood to form a rich humus that provides life and space for the next generation of trees and plants.

Effects of light

Wood reacts to strong light by a gradual change in colour. Some dark woods may become lighter, while lighter woods darken, even when protected with a clear finish. The wood is not harmed, and old pieces of furniture with this ageing effect, known as patina, are sought-after. Unfinished wood weathers to a grey colour when kept outside. Unless the wood is naturally durable or has been sealed or treated, surface breakdown can occur, leading to deterioration.

Fungi

Fungi are simple forms of plant life that feed on other living plants or plant material such as wood. Some are parasitic and use living hosts, others are saprophytic, extracting organic matter from dead wood.

Fungus spores grow in the presence of a food source that has a suitable level of moisture; they will not develop on wood dried below 20 per cent moisture content, in very low or high temperatures, or on wood that is saturated. Modifying or eliminating the various conditions will prevent infection, although the spores can lie dormant.

When spores land on damp wood, they germinate and put out fine branching filaments which multiply to form the mycelium, a mat that covers the surface or invades the wood to extract the nutrients.

Fungal attack

Some fungi, such as blue, or sap-stain, fungus which mainly attacks the sapwood of softwoods, cause discolouration of logs or sawn timber. Seasoning or fungicide treatment will control the infection.

Although fungi that attack cell structures, causing wood to decay or rot, are more serious, woodworkers have turned other fungal attacks to their advantage by utilizing infected light-coloured woods, such as maple and beech, for woodturning. Such 'spalted' wood displays patterns of black lines, containing patches of stained wood, both caused by the fungi. Similarly, beefsteak fungus attacks oak, leaving it a rich brown colour. The wood must be worked at the right time, to get the best markings while not allowing the wood to degrade to the point where it is friable and unworkable.

Fungi that invade and break down wood are known as wet and dry rots. They attack the cell structure, destroying the integrity of the wood and causing it to darken, shrink and split into cross-grain pieces. Wet rot occurs in wood with a high moisture content, and grows outdoors as well as indoors. Dry rot, so named for the desiccated, dried condition of affected wood, is particularly destructive to house timbers. It attacks damp wood in dark unventilated areas and seeks out wood to infect by sending out fine tubules, through which water is pumped to condition dry wood. Infected wood should be cut out and burnt, the surrounding structure should be treated with a fungicide, and damp conditions should be eliminated.

Insect attack

Insects can affect wood at any time. Wood-boring beetles attack both hardwoods and softwoods; the most common is the furniture beetle, also called woodworm. It will attack stored timber and household structural timbers, as well as items of furniture – unfinished wood and plywood drawer bottoms and backs are most at risk. A female beetle lays its eggs in crevices in the wood; these develop into larvae that burrow unnoticed through the wood, creating damaging tunnels as they feed. On maturity the adult beetles emerge through a small flight hole – the first sign of infestation – ready to repeat the reproductive cycle. Dark-coloured holes indicate an old outbreak, but fresh holes with fine dust or frass in and around the area are signs of active insects. Treatment with a chemical preservative will kill the larvae and prevent further outbreaks of infestation.

Other species that infest wood are the deathwatch and house long-horn beetles. These mainly attack house timbers, producing larger flight holes than those of the furniture beetle. Although rarer, they are potentially very destructive, and any outbreak should be reported to the appropriate local authority.

Because wood-boring beetles attack both growing and felled timber, all timber and veneers should be examined carefully before being selected for a project.

WOODS OF THE WORLD

CHAPTER 2 With practice, some common woods can be identified by their grain, colour, texture and smell. However, unfamiliar woods can be extremely difficult to identify, even by an expert. The following pages illustrate in colour a wide selection of commercially available woods from around the world.

Listings

The woods are listed alphabetically by the botanical classification of each genus and species. These are given in small type below the main entry, which is the standard commercial name. Other local or commercial names appear at the beginning of the text.

DISTRIBUTION OF TREES

The distribution of trees is primarily controlled by the prevailing climatic conditions. Rainfall, seasonal temperature, humidity, light and wind all play a part. Temperature is the most critical of these elements, and will determine the rate at which a species will grow, or if it will survive at all. This in turn is influenced by the local geographic environment; trees that grow well at low level would not necessarily survive at high altitude in the same latitude, but a species growing at high level in the south may well prosper in the lowlands of the cooler northern regions. Similarly, an east-to-west maritime or continental climate can influence the growing pattern of species as much as a north-to-south division. In addition to the seasonal and natural elements, others such as soil condition, local silviculture, and industrial and urban development, have an impact on the growth of trees. The world maps in this chapter illustrate in broad terms the distribution of predominant species of softwood and hardwood trees producing commercial timber (see pages 43 and 55).

World lumber-producing regions

The northern hemisphere is the prime source of the world's supply of commercial softwoods. Deciduous broadleaved hardwoods grow in the temperate northern hemisphere, and evergreen, mostly broadleaved, hardwoods in the tropics and southern hemisphere.

Softwoods

Most softwood trees are readily recognized by their excurrent form, having a single, tall, straight trunk with small lateral branches.

Excurrent form (Ponderosa pine)

Hardwoods

Most hardwood trees tend to be dendritic in form, having a trunk that divides and subdivides.

Dendritic form (American whitewood)

Converting a native oak
Traditional two-man hand-sawing techniques are used to cut to size a 17.5m (57ft) oak log in Berkshire. The sawn beam will be shaped with medieval-style tools to form the kcel for a replica of Christopher Columbus's ship, the *Santa Maria.*

SOFTWOODS OF THE WORLD

Softwood timber comes from cone-bearing, or coniferous, trees with exposed seeds, belonging to the botanical group *Gymnospermae*. It is this scientific grouping, rather than any purely physical properties, that gives softwoods their name.

When converted into boards, softwoods can be identified by their relatively light colour range, from pale yellow to reddish-brown. Other characteristic features are the grain pattern created by the change in colour and density of the earlywood and latewood (see page 13).

Cone-bearing trees
Although cone-bearing trees are mostly depicted as having a tall, pointed outline, this is not true of all conifers. Most are evergreens, with narrow, needle-shape leaves.

Softwood-producing regions of the world
The majority of the world's commercial softwoods come from countries in the northern hemisphere. These range from the Arctic and subarctic regions of Europe and North America down to the south-eastern United States.

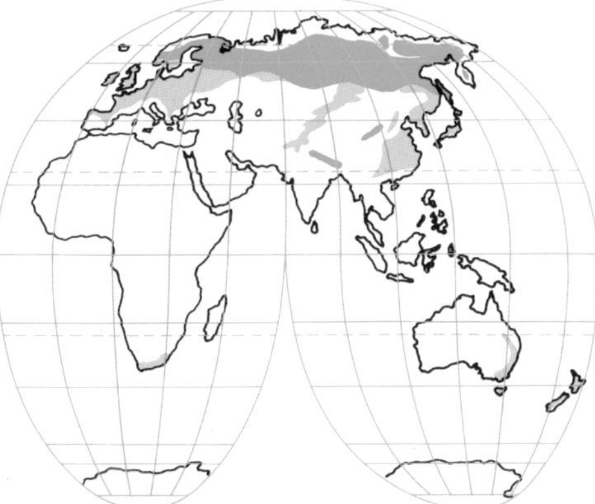

Distribution of softwoods
- Coniferous forest
- Mixed forest of coniferous and broadleaved deciduous

Cultivated softwoods
Grafting, cross-breeding and carefully controlled pollination are just some of the methods used today to produce fast-growing softwoods. The wood is cheaper than hardwoods, and is used in building construction, joinery and the manufacture of paper and fibreboard.

Buying softwood boards
Local sawmills will sell home-grown timber in whole boards. These can come complete with bark and waney edge, which is the uncut edge of the board. In contrast, imported boards are usually supplied de-barked or square-edged. Softwoods are generally sold as sawn boards or ready-planed; the latter may be referred to as planed-all-round (PAR) or planed-both-sides (PBS). The planing process removes at least 3mm (⅛in) from each side of the wood, so the actual size will be less than the nominal sawn size quoted by the timber merchant.

SILVER FIR

Abies alba

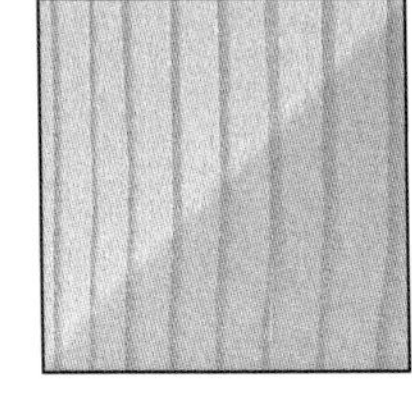

Other names: Whitewood.

Sources: Southern Europe, Central Europe.

Characteristics of the tree: A straight, thin tree, growing to about 40m (130ft) in height and 1m (3ft 3in) in diameter, losing its lower branches in the process.

Characteristics of the wood: The almost colourless, pale cream wood bears a resemblance to Norway spruce *(Picea abies)*, with straight grain and a fine texture. However, it is prone to knots and is not durable; for exterior uses, a preservative treatment is required.

Common uses: Building construction, joinery, plywood, boxes, poles.

Workability: It can be worked easily, using sharp hand and machine tools to produce a very smooth finish; it glues well.

Finishing: It takes stains, paints and varnishes readily.

Average dried weight: 480kg/m³ (30lb/ft³).

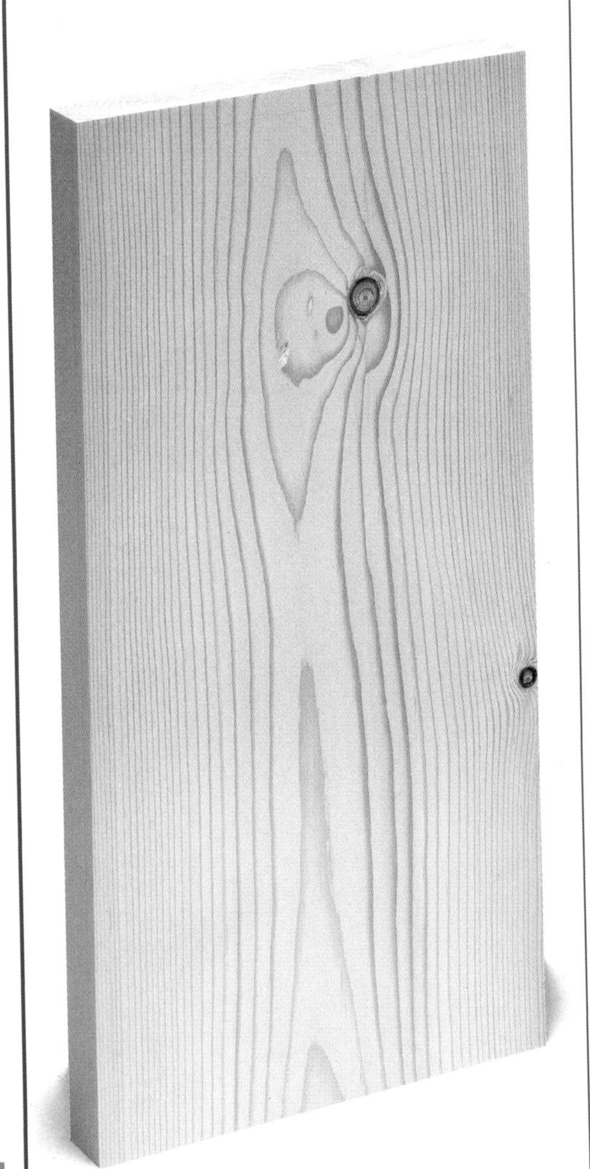

QUEENSLAND KAURI

Agathis spp.

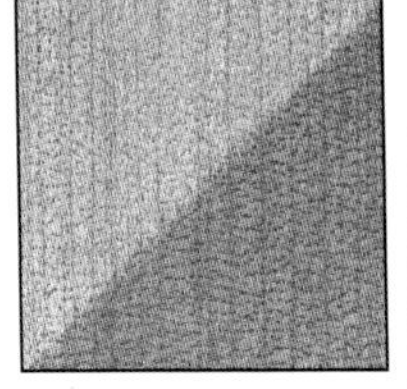

Other names: North Queensland kauri, South Queensland kauri.

Sources: Australia.

Characteristics of the tree: Although it can grow to more than 45m (150ft) high and 1.5m (5ft) in diameter, overcutting has led to a scarcity of larger trees; medium-size ones are the most common.

Characteristics of the wood: The straight-grained wood is not durable and varies in colour from pale cream-brown to pinkish-brown, with a fine, even texture and lustrous surface.

Common uses: Joinery, furniture.

Workability: It can be worked readily and brought to a fine, smooth finish using hand and machine tools; it glues well.

Finishing: It accepts stains and paints well, and can be polished to an excellent finish.

Average dried weight: 480kg/m³ (30lb/ft³).

PARANA PINE

Araucaria angustifolia

Other names: Brazilian pine (USA).

Sources: Brazil, Argentina, Paraguay.

Characteristics of the tree: It can reach about 36m (120ft) in height, with a flat crown of foliage at its top; the long, straight trunk can be up to 1m (3ft 3in) in diameter.

Characteristics of the wood: The mostly knot-free wood has barely perceptible growth rings, an even texture and straight grain; it is not durable, and must be well-seasoned to avoid large boards buckling. The core of the heartwood is dark brown, often flecked with streaks of bright red, while the rest is light brown.

Common uses: Joinery, furniture, plywood, turnery.

Workability: It is an easy wood to work and bring to a smooth finish with hand and machine tools; it glues well.

Finishing: It accepts paints, stains and polishes well.

Average dried weight: 530kg/m^3 (33lb/ft^3).

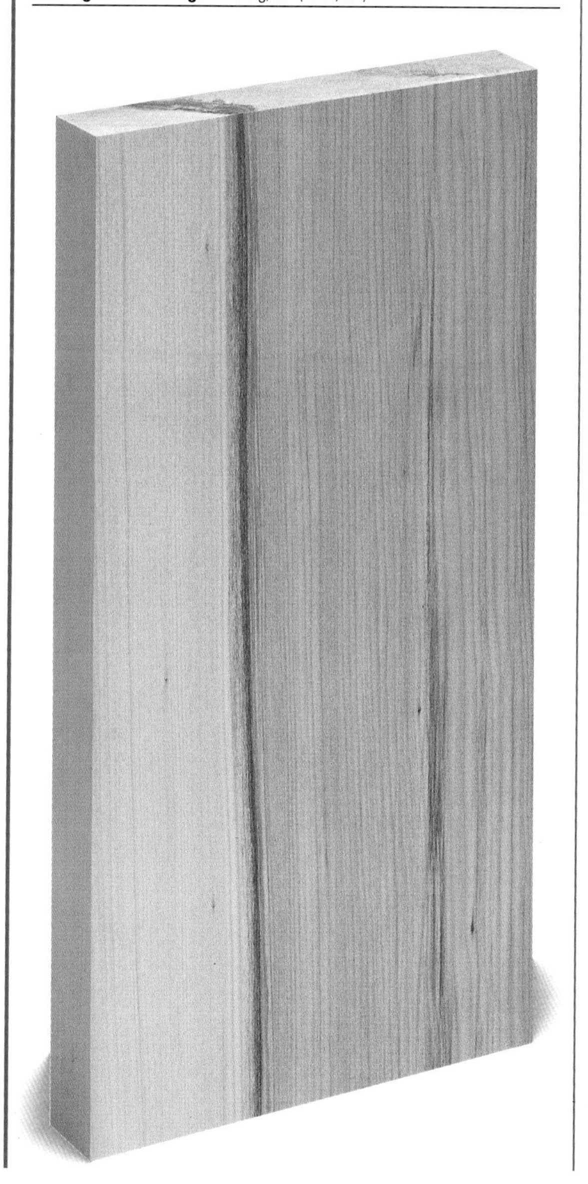

HOOP PINE

Araucaria cunninghamii

Other names: Queensland pine.

Sources: Australia, Papua New Guinea.

Characteristics of the tree: This tall, elegant tree, with tufts of foliage at the tips of thin branches, is not a true pine. The average height is about 30m (100ft), and the trunk diameter is about 1m (3ft 3in).

Characteristics of the wood: The versatile wood is not durable, and has straight grain and a fine texture; the heartwood is yellow-brown in colour, while the wide sapwood is light brown.

Common uses: Building construction, joinery, furniture, pattern-making, turnery, plywood.

Workability: If hand- and machine-tool cutting edges are kept sharp, to avoid tearing grain around fine knots, the wood can be worked easily; it also glues well.

Finishing: It accepts paints and stains well, and can be polished to an attractive finish.

Average dried weight: 560kg/m^3 (35lb/ft^3).

CEDAR OF LEBANON

Cedrus libani

Other names: True cedar.

Sources: Middle East.

Characteristics of the tree: Parkland-grown examples of this tree have large, low-growing branches and a distinctive broad crown of foliage. It can reach a height of about 40m (130ft) and a diameter of about 1.5m (5ft).

Characteristics of the wood: The aromatic wood is soft and durable, if brittle, with straight grain that is often clearly marked by the contrast between earlywood and latewood. The medium-fine-textured heartwood is light brown in colour.

Common uses: Building construction and joinery, interior and exterior furniture.

Workability: Although it can be worked easily with hand and machine tools and sands well, knots can be difficult to work.

Finishing: It accepts paints and stains well, and can be polished to a very fine finish.

Average dried weight: 560kg/m³ (35lb/ft³).

YELLOW CEDAR

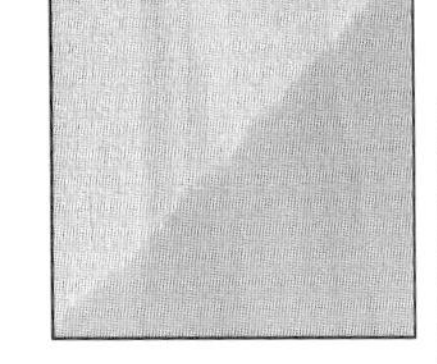

Chamaecyparis nootkatensis

Other names: Alaska yellow cedar, Pacific coast yellow cedar.

Sources: Pacific coast of North America.

Characteristics of the tree: This elegant, conical-shaped tree grows slowly to 30m (100ft) in height and about 1m (3ft 3in) in diameter.

Characteristics of the wood: The durable, pale yellow wood has straight grain and an even texture. When dry, it is relatively light, stiff, stable and very strong; it wears well and is resistant to decay.

Common uses: Furniture, veneers and high-class joinery – doors, windows, flooring, decorative panelling and mouldings – boatbuilding, oars, paddles.

Workability: It can be cut to fine tolerances, and glues well.

Finishing: It accepts paints and stains well, and can be polished to a fine finish.

Average dried weight: 500kg/m³ (31lb/ft³).

RIMU

Dacrydium cupressinum

Other names: Red pine.

Sources: New Zealand.

Characteristics of the tree: This tall, straight-growing tree can reach 36m (120ft) in height, and has a long, clean, straight trunk of up to 2.5m (8ft) in diameter.

Characteristics of the wood: The moderately durable wood has straight grain and a fine, even texture, with pale yellow sapwood that darkens to a reddish-brown heartwood. The colour of the somewhat indistinct figure, with patches and streaks of brown and yellow blending together, lightens and fades on exposure to light.

Common uses: Interior furniture, decorative veneer, turnery, panelling, plywood.

Workability: It can be worked well with hand and machine tools, and can be planed to a fine texture; it can be brought to a smooth finish. It glues well.

Finishing: It can be stained satisfactorily, and can be finished well with paints or polishes.

Average dried weight: 530kg/m³ (33lb/ft³).

LARCH

Larix decidua

Other names: None.

Sources: Europe, particularly mountainous areas.

Characteristics of the tree: One of the toughest softwoods, the larch grows to about 45m (150ft) in height, and the straight, cylindrical trunk has a diameter of about 1m (3ft 3in). The tree sheds its needles in winter.

Characteristics of the wood: The resinous wood is straight-grained and uniformly textured, and is relatively durable in outside use; however, hard knots can loosen after the seasoning process and can blunt cutting edges. The sapwood is narrow and light-coloured, while the heartwood is orange-red.

Common uses: Boat planking, pit props, joinery – including staircases, flooring, and door and window frames – posts, fencing.

Workability: The wood can be worked relatively easily with hand and machine tools; it sands well, although hard latewood grain may be left proud of the surface.

Finishing: It can be painted and varnished satisfactorily.

Average dried weight: 590kg/m³ (37lb/ft³).

NORWAY SPRUCE

Picea abies

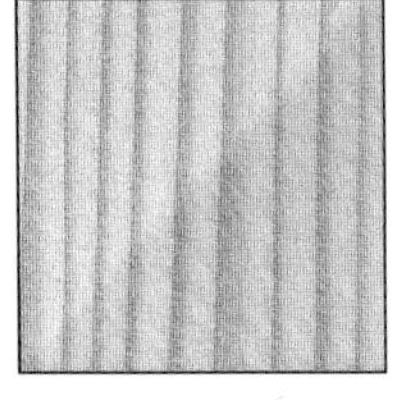

Other names: European whitewood, European spruce, whitewood.

Sources: Europe.

Characteristics of the tree: This important timber-producing tree has an average height of 36m (120ft), but can grow to 60m (200ft) in favourable conditions. Young trees are the source of the traditional Christmas tree.

Characteristics of the wood: The non-durable, lustrous wood is straight-grained and even-textured, with almost-white sapwood and pale yellow-brown heartwood. The strength properties are similar to the European redwood *(Pinus sylvestris)*, but with less prominent annual-growth rings.

Common uses: Interior building construction, flooring, boxes, plywood. Slow-grown wood is used for piano soundboards and the bellies of violins and guitars.

Workability: It can be worked easily with hand and machine tools, cuts cleanly and glues well.

Finishing: It accepts stains well, and can be finished satisfactorily with paints and varnishes.

Average dried weight: 450kg/m³ (28lb/ft³).

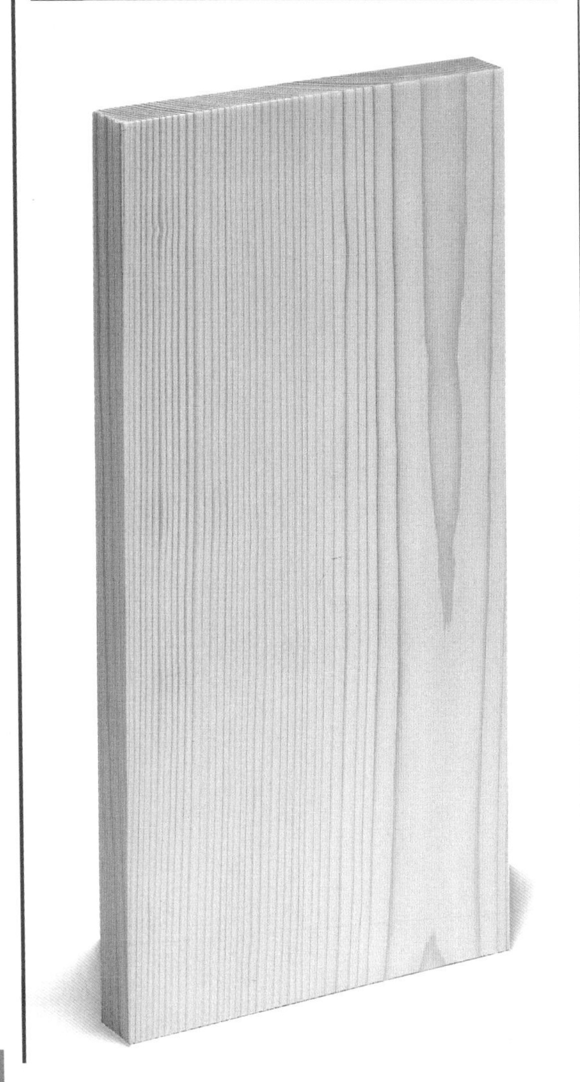

SITKA SPRUCE

Picea sitchensis

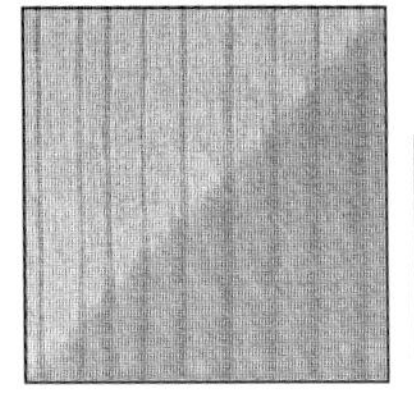

Other names: Silver spruce.

Sources: Canada, USA, UK.

Characteristics of the tree: This widely cultivated tree can reach a height of 87m (290ft), with a buttressed trunk of up to 5m (16ft) in diameter, but most fast-grown trees are smaller.

Characteristics of the wood: The non-durable wood is usually straight-grained and even-textured, with cream-white sapwood and slightly pink heartwood. It can be steam-bent, and is relatively light and strong, with good elasticity.

Common uses: Building construction, interior joinery, aircraft and gliders, boatbuilding, musical instruments, plywood.

Workability: It can be worked easily with hand and machine tools, but cutting edges must be kept sharp, to avoid tearing bands of earlywood. It glues well.

Finishing: It stains well, and can be finished satisfactorily with paints and varnishes.

Average dried weight: 450kg/m³ (28lb/ft³).

SUGAR PINE

Pinus lambertiana

Other names: Californian sugar pine.

Sources: USA.

Characteristics of the tree: It typically reaches about 45m (150ft) in height and 1m (3ft 3in) in diameter.

Characteristics of the wood: The even-grained wood is moderately soft, with a medium texture; it is not durable. The sapwood is white, and the heartwood a pale brown to reddish-brown colour.

Common uses: Light building construction, joinery.

Workability: Because of its softness, cutting edges must be kept sharp to avoid tearing the wood, but otherwise it can be worked well with hand and machine tools. It glues well.

Finishing: It can be brought to a satisfactory finish with stains, paints, varnishes and polishes.

Average dried weight: 420kg/m³ (26lb/ft³).

WESTERN
WHITE PINE

Pinus monticola

Other names: Idaho white pine.

Sources: USA, Canada.

Characteristics of the tree: The average height is 37m (125ft), and the straight trunk is about 1m (3ft 3in) in diameter.

Characteristics of the wood: The wood has straight grain and an even texture, and is not durable. Both earlywood and latewood are pale yellow to reddish-brown in colour, and the wood has fine resin-duct lines. It is similar to yellow pine *(Pinus strobus)* in many respects, but is tougher and shrinks slightly more.

Common uses: Building construction, joinery – including doors, windows and moulded skirting boards – boatbuilding, built-in furniture, pattern-making, plywood.

Workability: It is easily worked with hand and machine tools; it glues well.

Finishing: It accepts paints and varnishes well, and can be polished to a good finish.

Average dried weight: 450kg/m³ (28lb/ft³).

PONDEROSA PINE

Pinus ponderosa

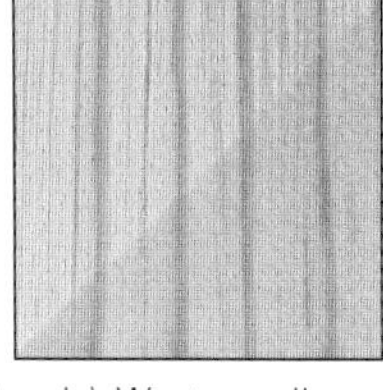

Other names: British Columbian soft pine (Canada); Western yellow pine, Californian white pine (USA).

Sources: USA, Canada.

Characteristics of the tree: This tree can reach 70m (230ft) in height; a typical straight trunk is about 750mm (2ft 6in) in diameter. It has an open, conical-shape crown.

Characteristics of the wood: The non-durable wood can be knotty and has resin ducts showing up as fine, dark lines on board surfaces. The wide, pale yellow sapwood is soft and even-textured; the heartwood is resinous, heavier, and a deep yellow to reddish-brown colour.

Common uses: Sapwood for pattern-making, doors, furniture, turnery; heartwood for joinery, building construction.

Workability: Both sapwood and heartwood can be worked well with hand and machine tools, but knots can cause problems when planing. It glues well.

Finishing: It takes paints and varnishes satisfactorily, but resinous wood needs a sealer before finishing.

Average dried weight: 480kg/m^3 (30lb/ft^3).

YELLOW PINE

Pinus strobus

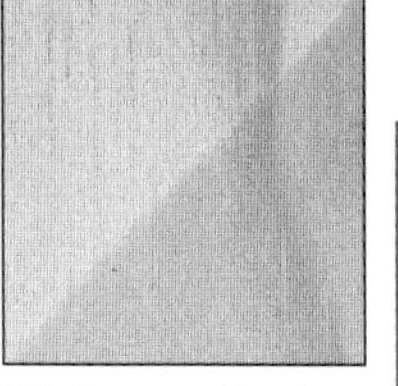

Other names: Quebec pine, Weymouth pine (UK); Eastern white pine, Northern white pine (USA).

Sources: USA, Canada.

Characteristics of the tree: It grows to about 30m (100ft) in height, and up to 1m (3ft 3in) in diameter.

Characteristics of the wood: Although the wood is soft, weak, and not durable, it is stable. It has straight grain, a fine, even texture, fine resin-duct marks and inconspicuous annual-growth rings; the colour varies from pale yellow to pale brown.

Common uses: High-class joinery, light building construction, furniture, engineering, pattern-making, carving.

Workability: It can be worked easily with hand and machine tools, if they are kept sharp; it glues well.

Finishing: It accepts stains, paints and varnishes, and polishes well.

Average dried weight: 420kg/m^3 (26lb/ft^3).

EUROPEAN
REDWOOD
Pinus sylvestris

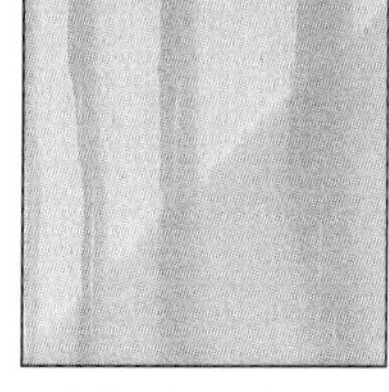

Other names: Scots pine, Scandinavian redwood, Russian redwood.

Sources: Europe, Northern Asia.

Characteristics of the tree: It grows up to 30m (100ft) in height and 1m (3ft 3in) in diameter. It is conical in shape when young, but becomes flat-topped when mature.

Characteristics of the wood: Although the resinous wood is stable and strong, it is not durable unless treated. The sapwood is a light white-yellow colour, and the heartwood varies from yellow-brown to reddish-brown; there is a distinct figure, with light earlywood and reddish latewood. The light colouring mellows with time.

Common uses: Building construction, interior joinery, turnery, plywood; selected knot-free timber is used for furniture.

Workability: Although knots and resin can cause problems, the wood works well with hand and machine tools, and glues well.

Finishing: It stains satisfactorily, but resin and latewood can prove resistant; it accepts paints and varnishes well, and can be polished to a good finish.

Average dried weight: 510kg/m³ (32lb/ft³).

DOUGLAS FIR
Pseudotsuga menziesii

Other names: British Columbian pine, Oregon pine.

Sources: Canada, Western USA, UK.

Characteristics of the tree: The average height is about 60m (200ft), but some trees reach 90m (300ft); trunks of forest-grown trees are up to 2m (6ft 6in) in diameter, and are free of branches for much of their height.

Characteristics of the wood: The straight-grained, reddish-brown wood is moderately durable, with distinctive earlywood and latewood grain, and produces large sizes of knot-free timber.

Common uses: Joinery, plywood, building construction.

Workability: It works well with hand and machine tools that have sharp cutting edges, glues satisfactorily, and can be finished smooth; however, latewood can be left proud of the surface after sanding.

Finishing: Latewood can be resistant to stains; earlywood takes them relatively well. Both accept paints and varnishes satisfactorily.

Average dried weight: 510kg/m³ (32lb/ft³).

SEQUOIA

Sequoia sempervirens

Other names: Californian redwood.

Sources: USA.

Characteristics of the tree: This magnificent, straight tree grows to about 100m (300ft) in height, and the buttressed trunk, with short, drooping branches, can exceed 4.5m (15ft) in diameter; even the distinctive, red-fissured bark can be more than 300mm (1ft) thick.

Characteristics of the wood: Despite being relatively soft, the straight-grained, reddish-brown wood is durable and suitable for exterior use. The texture can vary from fine and even to quite coarse, and there is a marked contrast between earlywood and latewood.

Common uses: Exterior cladding and shingles, interior joinery, coffins, posts.

Workability: As long as cutting edges are kept sharp to prevent break out along the cut, it can be worked well with hand and machine tools; it glues well.

Finishing: It sands and accepts paints well, and can be polished to a good finish.

Average dried weight: 420kg/m³ (26lb/ft³).

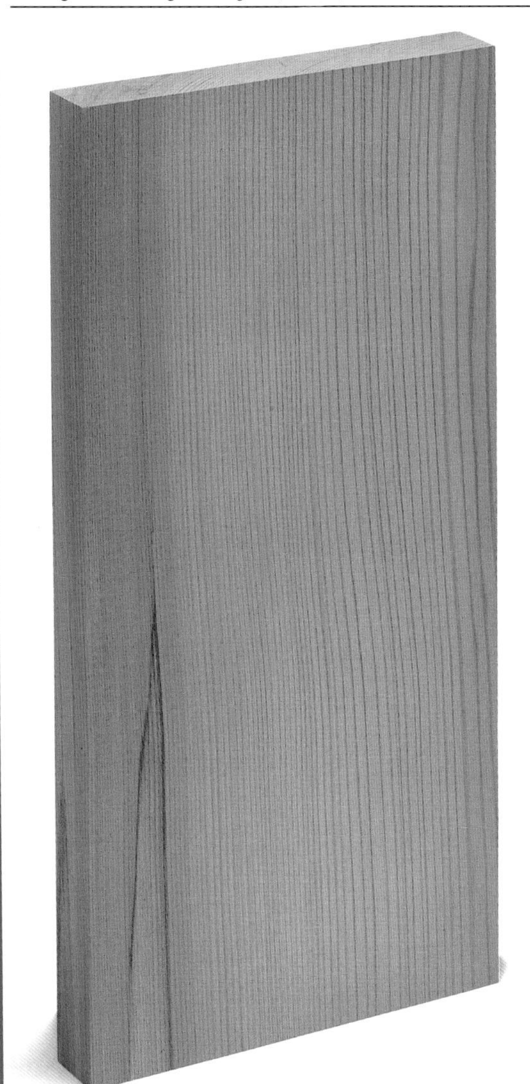

YEW

Taxus baccata

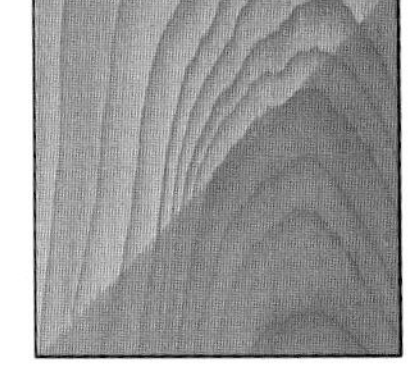

Other names: Common yew, European yew.

Sources: Europe, Asia Minor, North Africa, Myanmar, Himalayas.

Characteristics of the tree: The yew is the longest-living European tree – one specimen in Austria is over 3,500 years old. It grows to an average height of 15m (50ft), with dense, evergreen foliage and a short trunk of up to 6.1m (20ft) in diameter, deeply fluted where inter-grown shoots produce an irregular form.

Characteristics of the wood: The wood is hard, tough and durable, with a decorative growth pattern, orange-red heartwood and a distinct light-coloured sapwood, which often appears in irregular-shape boards, along with holes, small knots and bark inclusions. It is a good wood for steam-bending.

Common uses: Furniture, carving, interior joinery, veneer; it is particularly good for turning.

Workability: Straight-grained wood can be machined and hand-worked to a smooth finish, but irregular-grained wood can tear and be difficult to work. Its oily nature means care must be taken with gluing.

Finishing: It accepts stains satisfactorily and can be polished to an excellent finish.

Average dried weight: 670kg/m³ (42lb/ft³).

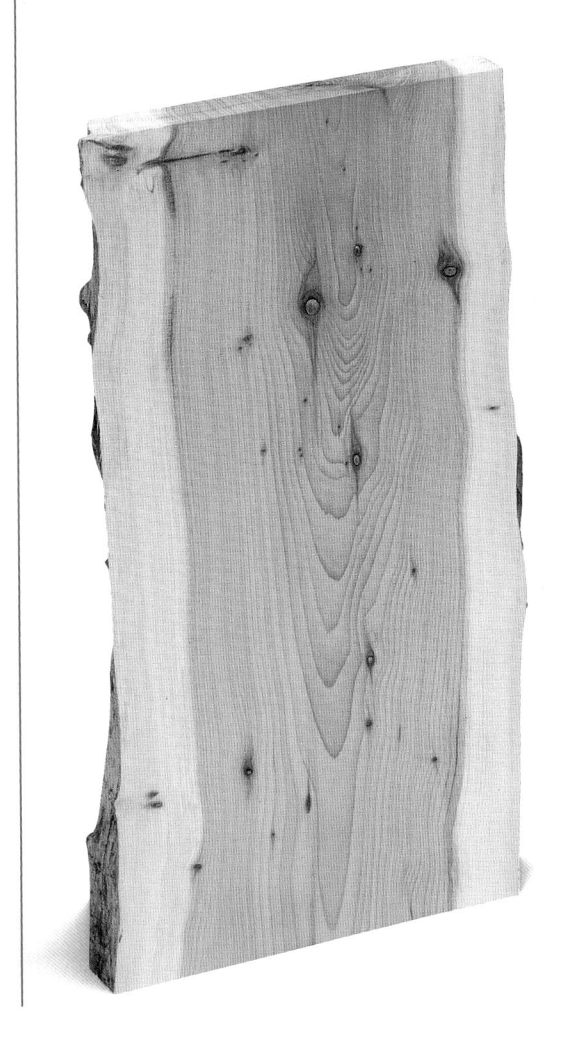

WESTERN
RED CEDAR
Thuja plicata

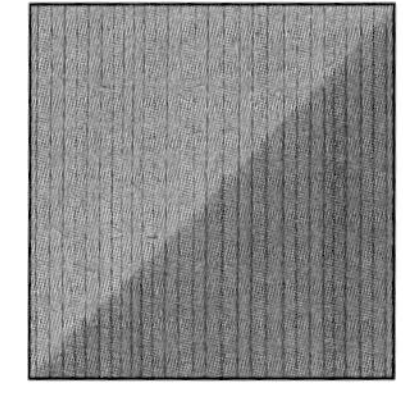

Other names: Giant arbor vitae (USA); red cedar (Canada); British Columbian red cedar (UK).

Sources: USA, Canada, UK, New Zealand.

Characteristics of the tree: This large, conically shaped, densely foliated tree reaches a height of up to 75m (250ft) and a diameter of up to 2.5m (8ft).

Characteristics of the wood: Although relatively soft and brittle, the non-resinous aromatic wood is durable; after long exposure to weathering, its reddish-brown colour fades to silver-grey. It has straight grain and a coarse texture.

Common uses: Joinery, shingles, exterior boarding, construction and furniture, cladding and decking, interior panelling.

Workability: It is easily worked with hand and machine tools, and glues well.

Finishing: It accepts paints and varnishes well, and can be brought to a good finish.

Average dried weight: 370kg/m³ (23lb/ft³).

WESTERN
HEMLOCK
Tsuga heterophylla

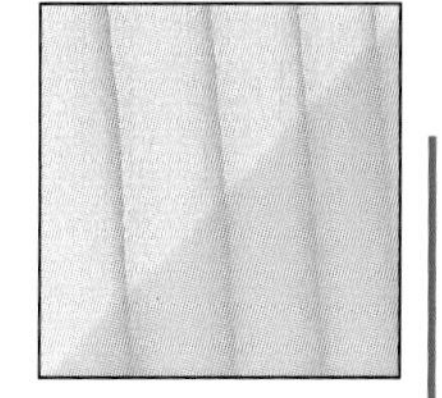

Other names: Pacific hemlock, British Columbian hemlock.

Sources: USA, Canada, UK.

Characteristics of the tree: This tall, straight, elegant tree with a distinctive drooping top can reach 60m (200ft) in height and 2m (6ft 6in) in diameter. It produces large pieces of timber.

Characteristics of the wood: The even-textured, straight-grained wood is not durable, and must be treated before exterior use; it is pale brown and semi-lustrous, knot-free and non-resinous, with relatively distinctive growth rings.

Common uses: Joinery, plywood, building construction, where it is often utilized in place of Douglas fir.

Workability: It can be worked easily with hand and machine tools, and glues well.

Finishing: It accepts stains, polishes, paints and varnishes well.

Average dried weight: 500kg/m³ (31lb/ft³).

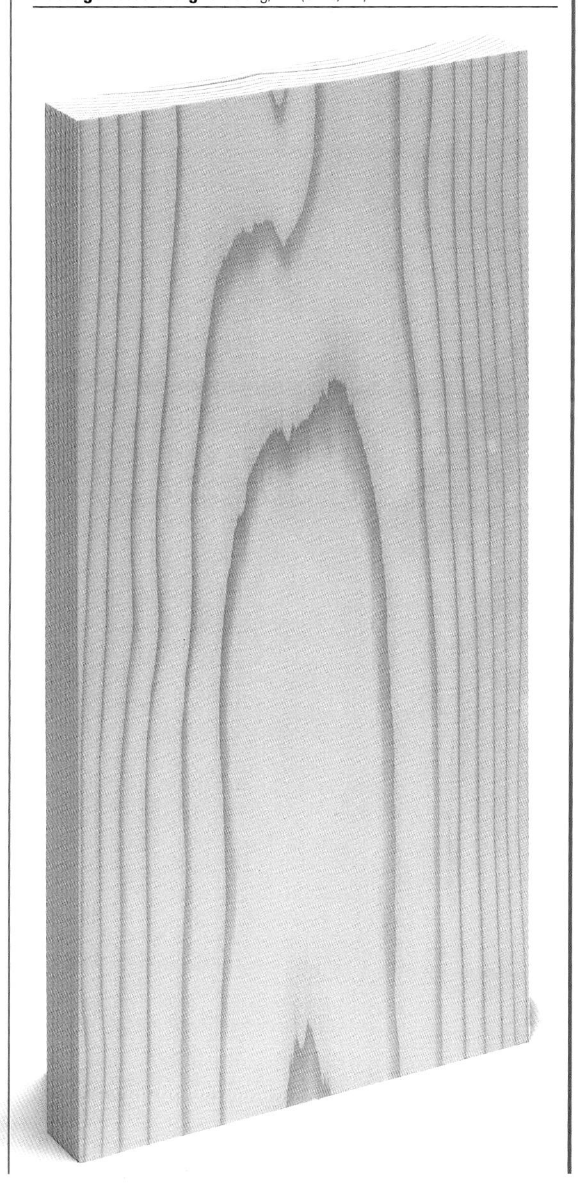

HARDWOODS OF THE WORLD

Hardwood trees belong to the botanical group *Angiospermae*, which are flowering broadleaved plants. Like softwoods, the term refers to the botanical grouping; however, it is true that most hardwoods are harder than softwood timbers. The greatest exception to this rule is balsawood; the tree belongs to the botanic hardwood group, yet has the softest timber commercially available from either group.

Of the thousands of species of hardwood trees that are found throughout the world, only a few hundred are harvested for commercial use.

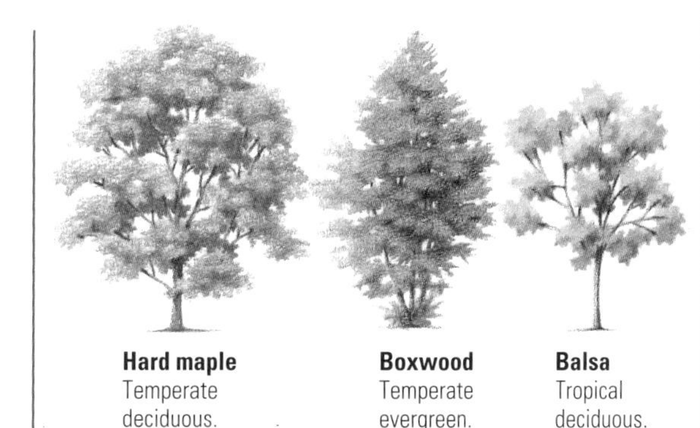

Hard maple Temperate deciduous.

Boxwood Temperate evergreen.

Balsa Tropical deciduous.

Broadleaved trees

Broadleaved trees grown in tropical forests are mainly evergreen. Most broadleaved trees grown in temperate zones are deciduous, losing their leaves in winter; some, however, have developed into evergreens.

Hardwood-producing regions of the world

Climate is the primary factor in determining where species grow. For the most part, deciduous broadleaved trees grow in the temperate northern hemisphere, and broadleaved evergreens are found in the southern hemisphere and tropical regions.

Distribution of hardwoods

- **Broadleaved evergreen forest**
- **Broadleaved deciduous forest**
- **Broadleaved evergreen and deciduous mixed forest**
- **Broadleaved deciduous and coniferous mixed forest**

Regenerating hardwood forests

Young trees are planted and tended in the natural environment of a tropical forest, to maintain hardwood stocks for future generations.

Hardwood veneers

Because hardwoods are generally more durable than softwoods and have a wider range of colour, texture and figure, they are sought-after and expensive. Highly prized, and increasingly rarer, exotic woods are often converted into veneer to satisfy demand (see page 88).

ENDANGERED HARDWOOD SPECIES

Over-production and a lack of international regulatory cooperation have led to a severe shortage of many tropical hardwoods *(see pages 14–16)**. In the following pages, those species marked with a red disc are at risk; check with suppliers that their woods come from certified sources.*

● **ENDANGERED**

AUSTRALIAN
BLACKWOOD
Acacia melanoxylon

Other names: Black wattle.

Sources: Mountainous regions of South and Eastern Australia.

Characteristics of the tree: It can reach 24m (80ft) in height, and has an average diameter of about 1.5m (5ft).

Characteristics of the wood: The strong wood is generally straight-grained, of a golden-brown to dark-brown colour, and has a medium, even texture. Interlocked, wavy grain is not uncommon, and selected timber produces an attractive fiddleback figure. The heartwood is durable and resistant to treatment with preservatives. It is a good wood for bending.

Common uses: Fine furniture, joinery, turnery, high-class fittings, decorative veneer.

Workability: It can be worked reasonably well using hand and machine tools, and straight-grained wood can be brought to a fine finish, although irregular grain can be difficult. It glues well.

Finishing: It accepts stains well, and polishes to a fine finish.

Average dried weight: 670kg/m^3 (41lb/ft^3).

EUROPEAN
SYCAMORE
Acer pseudoplatanus

Other names: Plane (Scotland); sycamore plane, great maple (UK).

Sources: Europe, Western Asia.

Characteristics of the tree: It is a moderate-size tree, reaching a height of 30m (100ft) and a diameter of 1.5m (5ft).

Characteristics of the wood: The lustrous white-to-yellowish-white wood is not durable and is unsuitable for exterior use; it is, however, good for steam-bending. It has a fine, even texture. Although it is generally straight-grained, wavy grain produces the fiddleback figure prized by musical-instrument-makers for violin backs.

Common uses: Turnery, furniture, flooring, veneer, kitchen utensils.

Workability: It can be worked well with hand and machine tools, if care is taken with wavy-grained wood, and glues well.

Finishing: It stains well and polishes to a fine finish.

Average dried weight: 630kg/m^3 (39lb/ft^3).

SOFT MAPLE

Acer rubrum

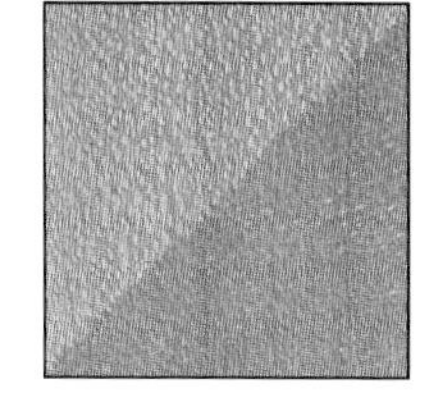

Other names: Red maple (USA, Canada).

Sources: USA, Canada.

Characteristics of the tree: This medium-size tree can reach 23m (75ft) in height and 750mm (2ft 6in) in diameter.

Characteristics of the wood: The light creamy-brown wood is straight-grained, with a lustrous surface and fine texture. It is not durable, and is not as strong as hard maple *(Acer saccharum)*, but is good for steam-bending.

Common uses: Furniture and interior joinery, musical instruments, flooring, turnery, plywood, veneer.

Workability: It works readily with hand and machine tools, and can be glued satisfactorily.

Finishing: It accepts stains well and can be polished to a fine finish.

Average dried weight: 630kg/m^3 (39lb/ft^3).

HARD MAPLE

Acer saccharum

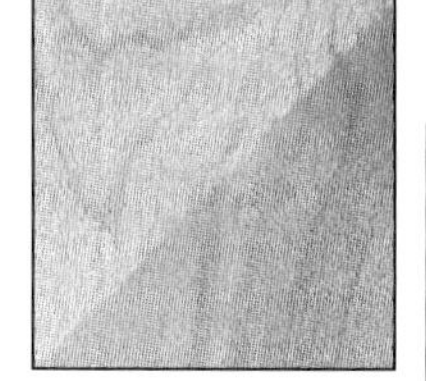

Other names: Rock maple, sugar maple.

Sources: Canada, USA.

Characteristics of the tree: It can grow to a moderate size of about 27m (90ft) in height and reach a diameter of 750mm (2ft 6in).

Characteristics of the wood: The heavy wood is hard-wearing but not durable, with straight grain and fine texture. The heartwood is light reddish-brown, while the light sapwood is often selected for its qualities of whiteness.

Common uses: Furniture, turnery, musical instruments, flooring, veneer, butcher's blocks.

Workability: The wood is difficult to work with hand or machine tools, particularly if irregularly grained; it glues well.

Finishing: It accepts stains and can be polished satisfactorily.

Average dried weight: 740kg/m^3 (46lb/ft^3).

RED ALDER

Alnus rubra

Other names: Western alder, Oregon alder.

Sources: Pacific coast of North America.

Characteristics of the tree: This smallish tree grows to a height of about 15m (50ft), with a diameter from 300 to 500mm (1ft to 1ft 8in).

Characteristics of the wood: The straight-grained, even-textured wood is soft and not particularly strong; it is not durable, but can be treated with preservative. The colour ranges from pale yellow to reddish-brown, with a subtle figure.

Common uses: Furniture, turnery, carving, decorative veneer, plywood, toy-making.

Workability: It can be worked well with hand and machine tools, if cutting edges are kept sharp, and glues well.

Finishing: It accepts stains well, and can be painted or polished to a fine finish.

Average dried weight: 530kg/m^3 (33lb/ft^3).

GONÇALO ALVES

Astronium fraxinifolium

Other names: Zebrawood (UK); tigerwood (USA).

Sources: Brazil.

Characteristics of the tree: The average height is 30m (100ft), but some trees reach 45m (150ft), and the diameter is about 1m (3ft 3in).

Characteristics of the wood: The medium-textured wood is hard and very durable, with hard and soft layers of material; its reddish-brown colour is streaked with dark brown, and is similar to rosewood. The grain is irregular and interlocked.

Common uses: Fine furniture, decorative woodware, turnery (for which it is particularly good), veneer.

Workability: It is a difficult wood to work by hand, and cutting edges of both hand and machine tools must be kept sharp. It has a natural lustre and glues well.

Finishing: It can be polished to a fine finish.

Average dried weight: 950kg/m^3 (59lb/ft^3).

ENDANGERED ●

YELLOW BIRCH

Betula alleghaniensis

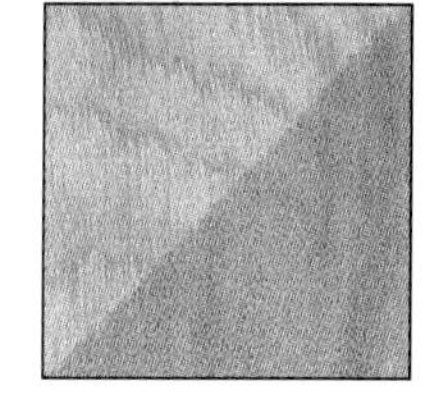

Other names: Hard birch, betula wood (Canada); Canadian yellow birch, Quebec birch, American birch (UK).

Sources: Canada, USA.

Characteristics of the tree: The largest North American birch usually reaches about 20m (65ft) in height, with a straight, slightly tapering trunk about 750mm (2ft 6in) in diameter.

Characteristics of the wood: The non-durable wood is usually straight-grained, has a fine, even texture, and is good for steam-bending. The light yellow sapwood is permeable, and the reddish-brown heartwood, with distinctive darker growth rings, is resistant to treatment with preservatives.

Common uses: Joinery, flooring, furniture, turnery, high-grade decorative plywood.

Workability: It can be worked reasonably well with hand tools, and well with machine tools; it glues well.

Finishing: It accepts stains well, and can be polished to a fine finish.

Average dried weight: 710kg/m^3 (44lb/ft^3).

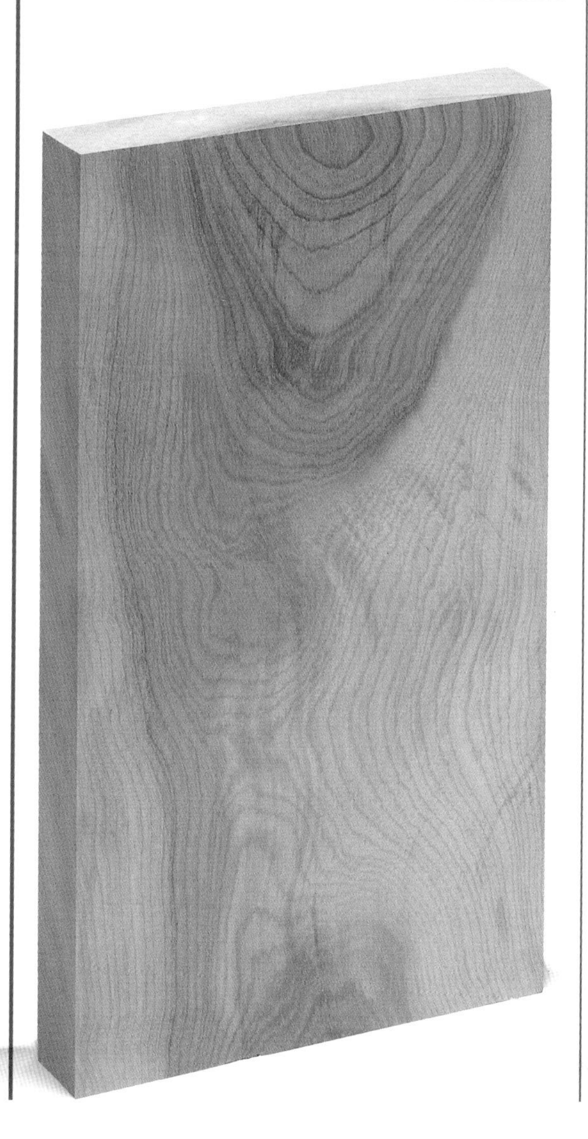

PAPER BIRCH

Betula papyrifera

Other names: American birch (UK); white birch (Canada).

Sources: USA, Canada.

Characteristics of the tree: The average height of this relatively small tree is about 18m (60ft), and the straight, clear, cylindrical trunk is about 300mm (1ft) in diameter.

Characteristics of the wood: The fairly hard wood has straight grain and a fine, even texture, and is moderately good for steam-bending; it is not durable. The sapwood is creamy white, and the heartwood, which is relatively resistant to treatment with preservatives, is pale brown.

Common uses: Turnery, domestic woodware and utensils, crates, plywood, veneer.

Workability: It can be worked reasonably well with hand and machine tools, and glues well.

Finishing: It accepts stains well, and can be polished to a fine finish.

Average dried weight: 640kg/m^3 (40lb/ft^3).

BOXWOOD

Buxus sempervirens

Other names: European, Turkish, Iranian boxwood, according to origin.

Sources: Southern Europe, Western Asia, Asia Minor.

Characteristics of the tree: This small, shrub-like tree reaches a height of up to 9m (30ft); the short lengths or billets produced are usually up to 1m (3ft 3in) long, with a diameter of up to 200mm (8in).

Characteristics of the wood: The wood is hard, tough, heavy and dense, with a fine, even texture and straight or irregular grain. When first cut it is pale yellow, but the colour mellows on exposure to light and air. The heartwood is durable, and the permeable sapwood can be treated with preservative. It has good steam-bending properties.

Common uses: Tool handles, engraving blocks, musical-instrument parts, rulers, inlay, turnery, carving.

Workability: Although it is a hard wood to work, sharp tools cut it very cleanly; it glues readily.

Finishing: It accepts stains well and polishes to a fine finish.

Average dried weight: 930kg/m³ (58lb/ft³).

SILKY OAK

Cardwellia sublimis

Other names: Bull oak, Australian silky oak (UK); Northern silky oak (Australia).

Sources: Australia.

Characteristics of the tree: It reaches a height of about 36m (120ft), and has a straight trunk up to 1.2m (4ft) in diameter.

Characteristics of the wood: The coarse, even-textured wood is reddish-brown in colour, with straight grain and large rays; it is moderately durable for exterior use, and is good for steam-bending, despite its moderate strength. Although a similar colour to American red oak *(Quercus rubra)*, it is not a true oak.

Common uses: Building construction, interior joinery, furniture, flooring, veneer.

Workability: It can be worked well with hand and machine tools; care must be taken not to tear the ray cells when planing. It glues well.

Finishing: It accepts stains well, and can be polished to a satisfactory finish.

Average dried weight: 550kg/m³ (34lb/ft³).

PECAN HICKORY

Carya illinoensis

Other names: Sweet pecan.

Sources: USA.

Characteristics of the tree: This edible-nut-bearing tree can reach a height of 30m (100ft) and a diameter of 1m (3ft 3in).

Characteristics of the wood: The dense, tough, coarse-textured wood is similar in appearance to ash *(Fraxinus* spp.*)*, with white sapwood and reddish-brown heartwood; it is shock-resistant, but not durable, and is excellent for steam-bending. The grain, though usually straight, can be irregular or wavy, and the growth rings are porous.

Common uses: Chairs and bentwood furniture, sports equipment, striking-tool handles, drumsticks.

Workability: If the tree has been grown fast, the dense wood will quickly dull cutting edges of hand and machine tools, making it difficult to work. It glues satisfactorily.

Finishing: It can be stained and polished well, despite the porosity.

Average dried weight: 750kg/m³ (46lb/ft³).

AMERICAN CHESTNUT

Castanea dentata

Other names: Wormy chestnut.

Sources: Canada, USA.

Characteristics of the tree: The average height is about 24m (80ft), and the average diameter about 500m (1ft 8in). Once plentiful, these trees were severely reduced in numbers early this century by chestnut blight, a fungal disease; many trees had to be felled to control the spread of the blight.

Characteristics of the wood: The durable wood is coarse-textured, with wide, prominent growth rings and light-brown heartwood. Like oak, it can develop a blue-black stain where it is in contact with ferrous metals, and corrodes them in damp conditions. The 'wormy' appearance is caused by insect attack, and is in demand for making reproduction 'period' furniture.

Common uses: Furniture, poles, stakes, coffins, veneer.

Workability: It is easy to work with hand and machine tools, and glues well.

Finishing: It accepts stains well, and can be polished to a fine finish.

Average dried weight: 480kg/m³ (30lb/ft³).

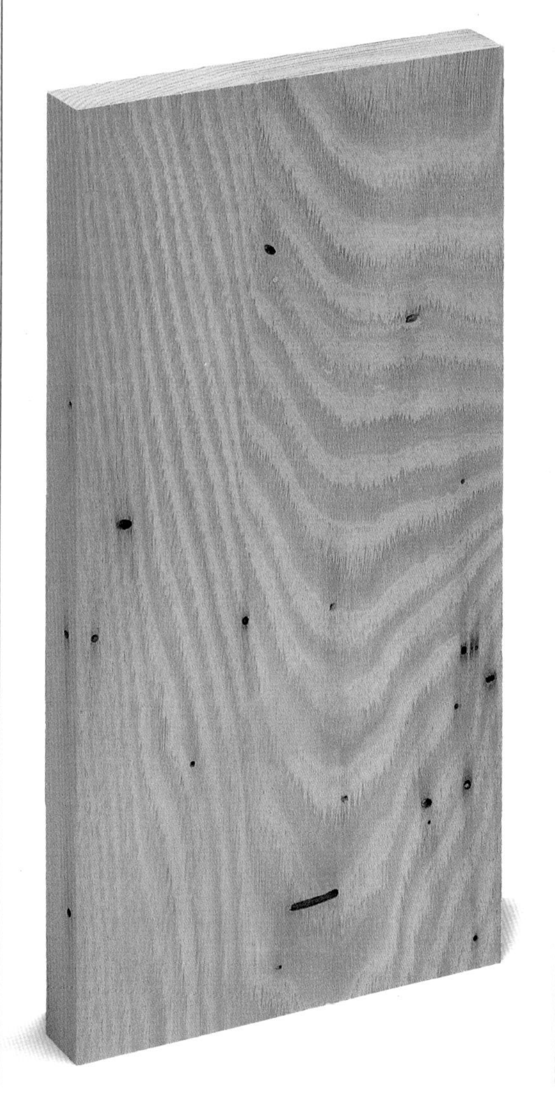

SWEET CHESTNUT

Castanea sativa

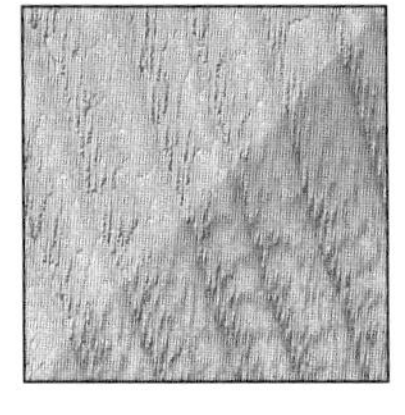

Other names: Spanish chestnut, European chestnut.

Sources: Europe, Asia Minor.

Characteristics of the tree: This largish, edible-nut-bearing tree can reach a height of more than 30m (100ft), and produces a straight trunk about 1.8m (6ft) in diameter and 6m (20ft) long.

Characteristics of the wood: The durable, coarse-textured wood is yellow-brown in colour, and has straight or spiral grain. When plain-sawn, the colour and texture resemble oak *(Quercus* spp.*)*; like oak and American chestnut *(Castanea dentata)*, the wood can corrode ferrous metals and become stained in contact with them.

Common uses: Furniture, turnery, coffins, poles, stakes.

Workability: It is easy to work with hand and machine tools, and the coarse texture can be brought to a smooth finish. It glues well.

Finishing: It accepts stains well, and can be varnished and polished to an excellent finish.

Average dried weight: 560kg/m^3 (35lb/ft^3).

BLACKBEAN

Castanospermum australe

Other names: Moreton Bay bean, Moreton Bay chestnut, beantree.

Sources: Eastern Australia.

Characteristics of the tree: This tall tree is found in moist forest regions from New South Wales to Queensland, and can reach about 40m (130ft) in height and 1m (3ft 3in) in diameter.

Characteristics of the wood: The hard, heavy wood is rich brown streaked with grey-brown; it has a rather coarse texture and an attractive figure. It is generally straight-grained, but interlocking grain is not uncommon; the heartwood is durable and resistant to treatment with preservatives.

Common uses: Furniture, turnery, joinery, carving, decorative veneers.

Workability: Because softer patches of this hard wood can crumble if cutting edges are not kept sharp, it is not particularly easy to work with hand or machine tools. In general, it glues reasonably well.

Finishing: It accepts stains well, and can be polished to a fine finish.

Average dried weight: 720kg/m^3 (45lb/ft^3).

SATINWOOD

Chloroxylon swietenia

Other names: East Indian satinwood.

Sources: Central and Southern India, Sri Lanka.

Characteristics of the tree: This smallish tree reaches about 15m (50ft) in height, with a straight trunk about 300mm (1ft) in diameter.

Characteristics of the wood: The lustrous, durable wood is light yellow to golden-brown in colour, with a fine, even texture and interlocked grain that produces a striped figure. It is heavy, hard and strong.

Common uses: Interior joinery, furniture, veneer, inlay, turnery.

Workability: It is a moderately difficult wood to work with hand and machine tools, and to glue.

Finishing: If care is taken, it can be brought to a smooth surface and polished to a fine finish.

Average dried weight: 990kg/m^3 (61lb/ft^3).

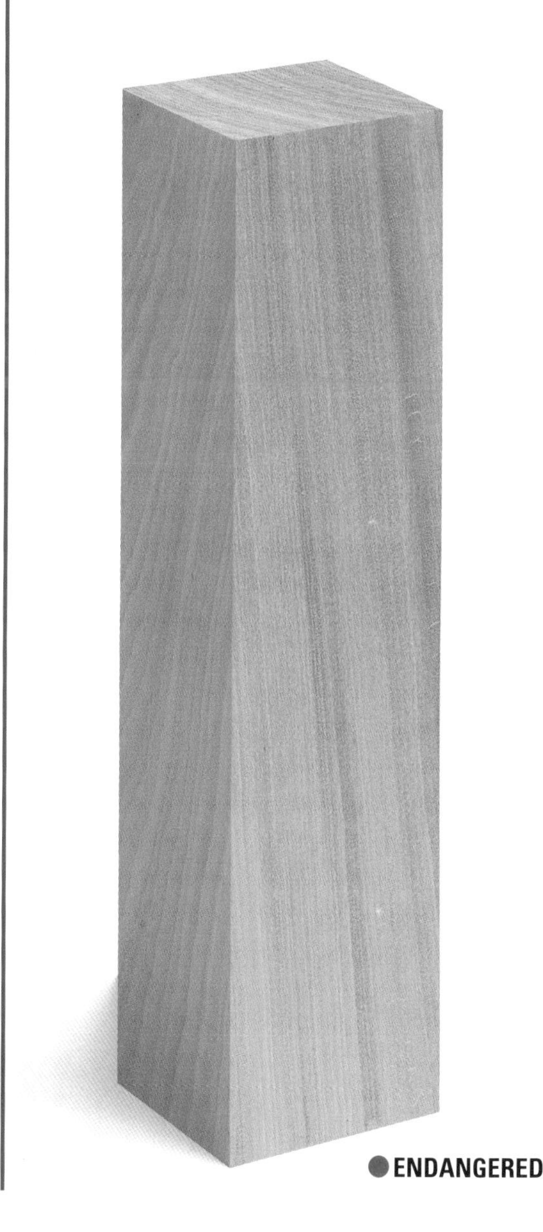

● ENDANGERED

KINGWOOD

Dalbergia cearensis

Other names: Violet wood, violetta (USA); bois violet (France); violete (Brazil).

Sources: South America.

Characteristics of the tree: This small tree, botanically related to rosewood, produces short logs or billets of wood up to 2.5m (8ft) long; with the white sapwood removed, the diameter of the billets is between 75 and 200mm (3 and 8in).

Characteristics of the wood: The fine, even-textured, durable wood is usually straight-grained, and the dark, lustrous heartwood has a variegated striped violet-brown, black and golden-yellow figure.

Common uses: Turnery, inlay, marquetry.

Workability: If cutting edges are kept sharp, it is an easy wood to work; it glues satisfactorily.

Finishing: It can be burnished to a fine finish, and can be polished well with wax.

Average dried weight: 1200kg/m^3 (75lb/ft^3).

● ENDANGERED

INDIAN ROSEWOOD

Dalbergia latifolia

Other names: East Indian rosewood, Bombay rosewood (UK); Bombay blackwood (India).

Sources: India.

Characteristics of the tree: This tree can reach 24m (80ft) in height; the straight, clear, cylindrical trunk can be up to 1.5m (5ft) in diameter.

Characteristics of the wood: The durable wood is hard, heavy, and has a uniform, moderately coarse texture; the colour is golden-brown to purple-brown, streaked with black or dark purple, and interlocked narrow bands of grain produce a subtle ribbon figure.

Common uses: Furniture, musical instruments, boatbuilding, shop fittings, turnery, veneer.

Workability: It is moderately difficult to work using hand tools, and dulls the cutting edges of machine tools. It glues satisfactorily.

Finishing: Although the grain requires filling in order to achieve a high polish, it can be finished well with wax.

Average dried weight: 870kg/m^3 (54lb/ft^3).

● **ENDANGERED**

COCOBOLO

Dalbergia retusa

Other names: Granadillo (Mexico).

Sources: West coast of Central America.

Characteristics of the tree: This moderate-size tree can reach a height of 30m (100ft), and can produce a fluted trunk about 1m (3ft 3in) in diameter.

Characteristics of the wood: The durable, irregular-grained wood is hard and heavy, with a uniform medium-fine texture. The heartwood has a variegated colour, from purple-red to yellow, with black markings; on exposure, the colour turns to deep orange-red.

Common uses: Turnery, brush backs, cutlery handles, veneer.

Workability: Although hard, it can be worked readily with hand and machine tools, as long as the cutting edges are kept sharp. The oily nature of the wood means it can be machined to a fine, smooth surface; it is, however, difficult to glue.

Finishing: It can be stained and polished to a fine finish.

Average dried weight: 1100kg/m^3 (68lb/ft^3).

● **ENDANGERED**

EBONY

Diospyros ebenum

Other names: Tendo, tuki, ebans.

Sources: Sri Lanka, India.

Characteristics of the tree: This relatively small tree grows up to 30m (100ft) in height, of which the straight trunk is about 4.5m (15ft), and about 750mm (2ft 6in) in diameter.

Characteristics of the wood: The hard, heavy and dense wood can have straight, irregular or wavy grain, and has a fine even texture. The non-durable sapwood is yellowish-white, and the durable, lustrous heartwood is the familiar dark brown to black.

Common uses: Turnery, musical instruments, inlay, cutlery handles.

Workability: Other than on the lathe, it is a difficult wood to work, because it tends to chip and to dull cutting edges quickly. It does not glue well.

Finishing: It can be polished to an excellent finish.

Average dried weight: 1190kg/m³ (74lb/ft³).

JELUTONG

Dyera costulata

Other names: Jelutong bukit, jelutong paya (Sarawak).

Sources: Southeast Asia.

Characteristics of the tree: This large tree can reach a height of 60m (200ft), and the long, straight trunk can grow to 27m (90ft) high, with a diameter of up to 2.5m (8ft).

Characteristics of the wood: The soft, straight-grained wood has a lustrous, fine and even texture, and a plainish figure; it is not durable. There are usually latex ducts about 12mm (½in) wide. Both the sapwood and heartwood are a creamy, pale brown colour.

Common uses: Interior joinery, pattern-making, matches, plywood.

Workability: It can be worked easily and brought to a smooth finish with hand and machine tools, and it is easy to carve. It glues well.

Finishing: It accepts stains and varnishes well, and can be polished to a fine finish.

Average dried weight: 470kg/m³ (29lb/ft³).

● ENDANGERED

● ENDANGERED

QUEENSLAND WALNUT

Endiandra palmerstonii

Other names: Australian walnut, walnut bean, oriental wood.

Sources: Australia.

Characteristics of the tree: This tall tree can reach a height of 42m (140ft), and the long, buttressed trunk is about 1.5m (5ft) in diameter.

Characteristics of the wood: Although the non-durable wood looks similar to that of the European walnut *(Juglans regia)*, it is not a true walnut. The colour can vary from light to dark brown, streaked with pink and dark grey; the interlocked and wavy grain produces an attractive figure. The ray cells often contain silica.

Common uses: Furniture, interior joinery, shop fittings, flooring, decorative veneer.

Workability: It is a difficult wood to work with hand or machine tools, due to its dulling effect on cutting edges, but it can be brought to a smooth natural finish and glues satisfactorily.

Finishing: It polishes to a fine finish.

Average dried weight: 690kg/m^3 (43lb/ft^3).

UTILE

Etandrophragma utile

Other names: Sipo (Ivory Coast); assié (Cameroon).

Sources: Africa.

Characteristics of the tree: It is a tall tree, about 45m (150ft) in height, with a straight cylindrical trunk about 2m (6ft 6in) in diameter.

Characteristics of the wood: The moderately strong, durable wood is pinkish-brown when freshly cut, deepening with exposure to reddish-brown. It has a medium texture, and the interlocked grain produces a ribbon-stripe figure when quarter-sawn.

Common uses: Interior and exterior joinery, boatbuilding, furniture, flooring, plywood, veneer.

Workability: If care is taken not to tear the ribbon-stripe figure when planing, the wood can be worked well with hand and machine tools; it glues well.

Finishing: It accepts stains and polishes well.

Average dried weight: 660kg/m^3 (41lb/ft^3).

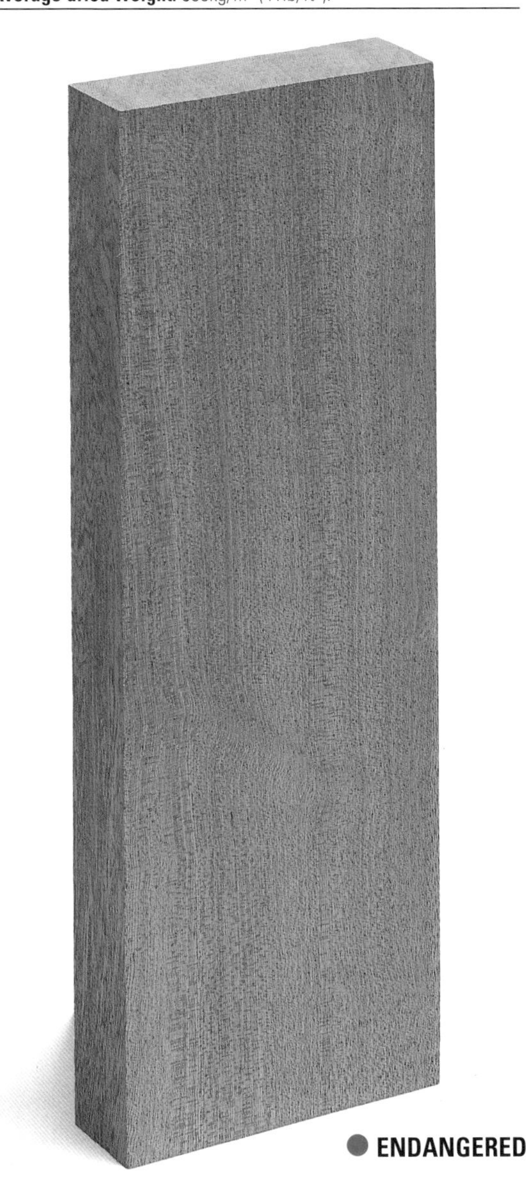

● **ENDANGERED**

JARRAH

Eucalyptus marginata

Other names: None.

Sources: Western Australia.

Characteristics of the tree: This tall tree can reach a height of 45m (150ft), with a long, clear trunk about 1.5m (5ft) in diameter.

Characteristics of the wood: The very durable wood is strong, hard and heavy, with an even, medium-coarse texture. The narrow sapwood is a yellowish-white colour, while the heartwood, light to dark red when first cut, deepens to red-brown. The grain is usually straight, but can also be wavy or interlocking; the figure displays fine brown decorative flecks, caused by the fungus *Fistulina hepatica*, and occasional gum veins.

Common uses: Building and marine construction, exterior and interior joinery, furniture, turnery, decorative veneers.

Workability: Although moderately difficult to work with hand or machine tools, it is good for turning; it glues well.

Finishing: It polishes very well, particularly with an oil finish.

Average dried weight: 820kg/m^3 (51lb/ft^3).

AMERICAN BEECH

Fagus grandifolia

Other names: None.

Sources: Canada, USA.

Characteristics of the tree: This relatively small tree reaches an average height of 15m (50ft), and produces a trunk of about 500mm (1ft 8in) in diameter.

Characteristics of the wood: Slightly coarser and heavier than European beech *(Fagus sylvatica)*, this straight-grained wood has similar strength and good steam-bending properties. It is light brown to reddish-brown in colour, with a fine, even texture; although it is perishable on exposure to moisture, it can be treated with preservative successfully.

Common uses: Cabinetmaking, interior joinery, turnery, bentwood furniture.

Workability: It can be worked well with hand and machine tools, though it has a propensity to scorch on crosscutting and drilling. It glues well.

Finishing: It accepts stain well, and can be polished to a fine finish.

Average dried weight: 740kg/m^3 (46lb/ft^3).

EUROPEAN BEECH

Fagus sylvatica

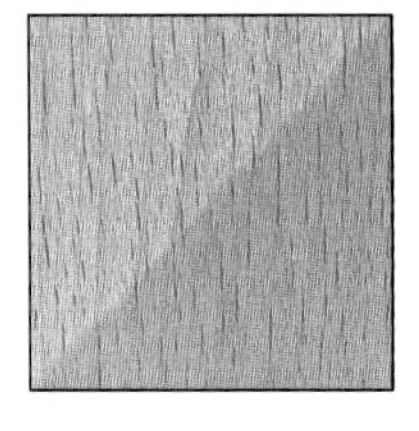

Other names: English, French, Danish beech etc., according to origin.

Sources: Europe.

Characteristics of the tree: This large tree can reach a height of 45m (150ft), with a straight, clear trunk about 1.2m (4ft) in diameter.

Characteristics of the wood: When first cut, the straight-grained, fine, even-textured wood is whitish-brown, and deepens to yellowish-brown on exposure; 'steamed beech', wood that has been steamed as part of the seasoning process, is a reddish-brown. It is a strong wood, excellent for steam-bending and, when seasoned, is tougher than oak. Although perishable, it can be treated with preservative.

Common uses: Interior joinery, cabinetmaking, turnery, bentwood furniture, plywood, veneer.

Workability: It works readily with hand and machine tools, but the ease of working depends on the quality and seasoning; it glues well.

Finishing: It accepts stains well, and can be polished to a fine finish.

Average dried weight: 720kg/m^3 (45lb/ft^3).

AMERICAN
WHITE ASH

Fraxinus americana

Other names: Canadian ash (UK); white ash (USA).

Sources: Canada, USA.

Characteristics of the tree: This tree grows up to about 18m (60ft) in height, and has a trunk diameter of about 750mm (2ft 6in).

Characteristics of the wood: The strong, shock-resistant wood is ring-porous, with a distinct figure. It has coarse and generally straight grain, with an almost white sapwood and pale brown heartwood, similar to European ash. Although non-durable, treatment with preservative allows exterior use; it is a good wood for steam-bending.

Common uses: Joinery, boatbuilding, sports equipment, tool handles, plywood, veneer.

Workability: It works well with hand and machine tools, and can be brought to a fine surface finish; it glues well.

Finishing: It accepts stains well, is often finished in black, and can be polished to a fine finish.

Average dried weight: 670kg/m^3 (42lb/ft^3).

EUROPEAN ASH

Fraxinus excelsior

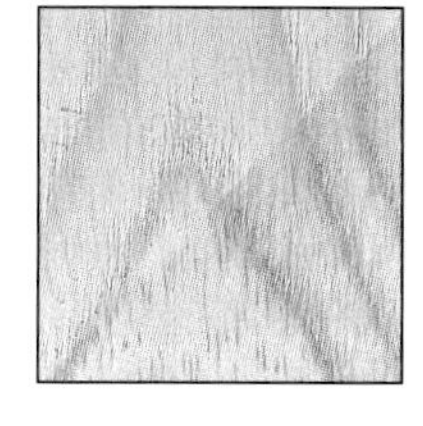

Other names: English, French, Polish ash etc., according to origin.

Sources: Europe.

Characteristics of the tree: It is a moderate-to-large tree, averaging a height of 30m (100ft), and with a trunk from 500mm to 1.5m (1ft 8in to 5ft) in diameter.

Characteristics of the wood: The tough, coarse-textured, straight-grained wood is flexible and relatively split- and shock-resistant, and is excellent for steam-bending. Both sapwood and heartwood are whitish to pale brown; 'olive ash' is produced from logs with dark-stained heartwood, and pale, strong 'sports ash' is in high demand. The wood is perishable, and is only suitable for exterior use when it has been treated with preservative.

Common uses: Sports equipment and tool handles, cabinetmaking, bentwood furniture, boatbuilding, vehicle bodies, ladder rungs, laminated work, plywood, decorative veneer.

Workability: It can be worked well with hand and machine tools, and can be brought to a fine surface finish; it glues well.

Finishing: It accepts stains well, and can be polished to a fine finish.

Average dried weight: 710kg/m^3 (44lb/ft^3).

RAMIN

Gonystylus macrophyllum

Other names: Melawis (Malaysia); ramin telur (Sarawak).

Sources: Southeast Asia.

Characteristics of the tree: It reaches a height of about 24m (80ft), and has a long, straight trunk about 600mm (2ft) in diameter.

Characteristics of the wood: The moderately fine, even-textured wood usually has straight grain, but this can be slightly interlocked. Both sapwood and heartwood are a pale cream-brown colour. The wood is perishable, and is not suited for exterior use.

Common uses: Interior joinery, flooring, furniture, toy-making, turnery, carving, veneer.

Workability: It can be worked reasonably well with hand and machine tools, but care must be taken to keep cutting edges sharp. It glues well.

Finishing: It accepts stains, paints and varnishes well, and can be polished to a satisfactory finish.

Average dried weight: 670kg/m^3 (41lb/ft^3).

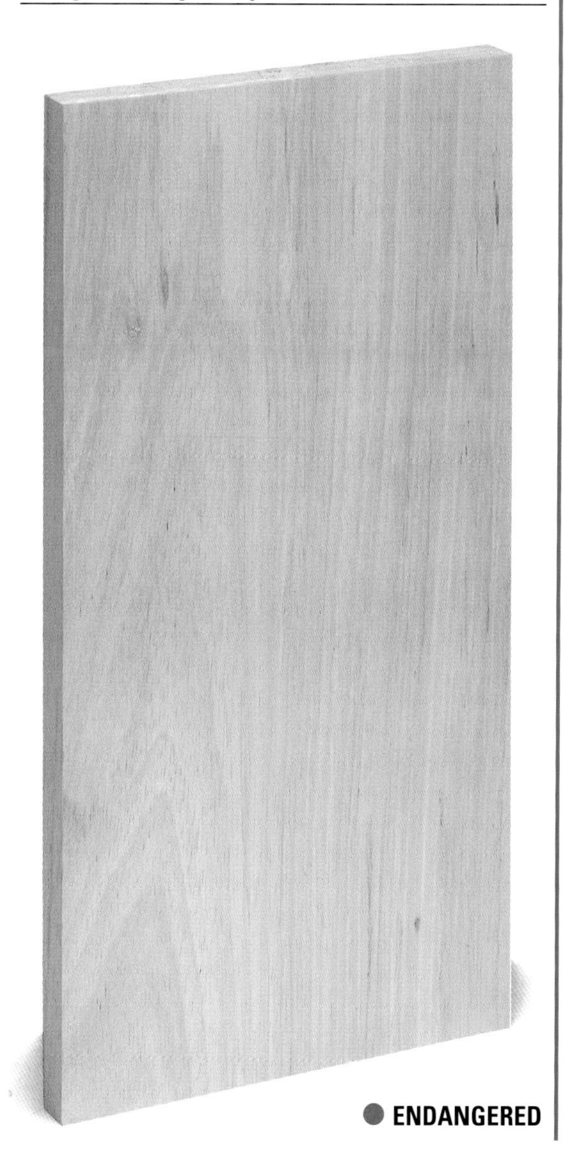

● **ENDANGERED**

LIGNUM VITAE

Guaiacum officinale

Other names: Ironwood (USA); bois de gaiac (France); guayacan (Spain); pala santo, guayacan negro (Cuba).

Sources: West Indies, tropical America.

Characteristics of the tree: This small tree grows slowly to 9m (30ft) in height, with a diameter of about 500mm (1ft 8in); the wood is sold in short billets.

Characteristics of the wood: The fine, uniform-textured wood, with closely interlocked grain, is one of the hardest and heaviest commercial timbers. It is very durable and resinous, with an oily feel; its hardness and self-lubricating properties are much in demand. The narrow sapwood is cream-coloured, and the heartwood is dark greenish-brown to black.

Common uses: Bearings and pulleys, mallets, turnery.

Workability: It is very difficult to saw and work with hand and machine tools, but can be brought to a fine finish on a lathe. An oil solvent must be used for it to glue well.

Finishing: It can be burnished to a fine natural finish.

Average dried weight: 1250kg/m^3 (78lb/ft^3).

● **ENDANGERED**

Lignum vitae is listed in Appendix II in CITES (see page 14).

BUBINGA

Guibourtia demeusei

Other names: African rosewood; kevazingo (Gabon); essingang (Cameroon).

Sources: Cameroon, Gabon, Zaire.

Characteristics of the tree: It grows to about 30m (100ft) in height, and the long, straight trunk reaches a diameter of about 1m (3ft 3in).

Characteristics of the wood: The hard, heavy wood has a moderately coarse, even texture; although not resilient, it is reasonably durable and strong. The grain can be straight or interlocking and irregular, and the heartwood is red-brown, with red and purple veining.

Common uses: Furniture, woodware, turnery, decorative veneer (known as kevazingo when rotary cut).

Workability: Although it can be worked well with hand tools and machined to a fine finish, cutting edges must be kept sharp. Gum pockets in the wood can cause problems when gluing.

Finishing: It accepts stains well and can be polished to a fine finish.

Average dried weight: 880kg/m^3 (55lb/ft^3).

● **ENDANGERED**

BRAZILWOOD

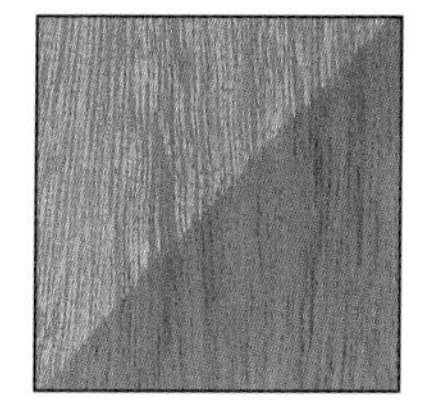

Guilandina echinata

Other names: Pernambuco wood, bahia wood, para wood.

Sources: Brazil.

Characteristics of the tree: This small-to-medium-size tree produces short billets or lengths up to 200mm (8in) in diameter.

Characteristics of the wood: The heavy, hard wood is tough, resilient and very durable, with generally straight grain and a fine, even texture. The sapwood is pale, in contrast to the heartwood, which is a lustrous bright orange-red that turns to a rich red-brown on exposure.

Common uses: Dyewood, violin bows, exterior joinery, parquet flooring, turnery, gun stocks, veneer.

Workability: It can be worked reasonably well with hand and machine tools, as long as cutting edges are kept sharp, and glues well.

Finishing: The surface can be polished to an exceptionally fine finish.

Average dried weight: 1280kg/m³ (80lb/ft³).

● **ENDANGERED**

BUTTERNUT

Juglans cinerea

Other names: White walnut.

Sources: Canada, USA.

Characteristics of the tree: This relatively small tree can reach a height of about 15m (50ft), and can have a trunk diameter of about 750mm (2ft 6in).

Characteristics of the wood: The straight-grained, coarse-textured wood is relatively soft and weak, and is not durable. The figure resembles that of black American walnut *(Juglans nigra)*, but the heartwood, from medium brown to dark brown, is lighter in colour.

Common uses: Furniture, interior joinery, carving, veneer, boxes, crates.

Workability: With sharp cutting edges, it can be worked easily with hand and machine tools. It glues well.

Finishing: It accepts stains well and can be polished to a fine finish.

Average dried weight: 450kg/m³ (28lb/ft³).

AMERICAN WALNUT

Juglans nigra

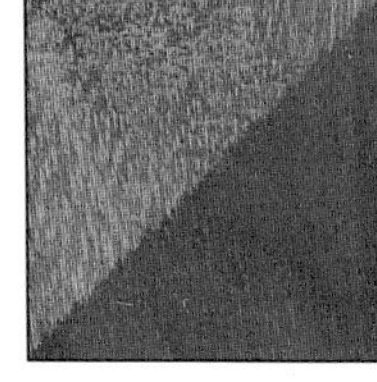

Other names: Black American walnut.

Sources: USA, Canada.

Characteristics of the tree: It grows to a height of about 30m (100ft), with a trunk about 1.5m (5ft) in diameter.

Characteristics of the wood: The tough, moderately durable wood has an even but coarse texture; the grain is usually straight, but can be wavy. Light-coloured sapwood contrasts with rich, dark brown-purplish heartwood. It is a good wood for steam-bending.

Common uses: Furniture, musical instruments, interior joinery, gun stocks, turnery, carving, plywood, veneer.

Workability: It can be worked well with hand and machine tools; it glues well.

Finishing: It can be polished to a fine finish.

Average dried weight: 660kg/m^3 (41lb/ft^3).

EUROPEAN WALNUT

Juglans regia

Other names: English, French, Italian walnut etc., according to origin.

Sources: Europe, Asia Minor, Southwest Asia.

Characteristics of the tree: This nut-bearing tree reaches a height of about 30m (100ft), and the average trunk diameter is 1m (3ft 3in).

Characteristics of the wood: The moderately durable wood has a rather coarse texture; the straight-to-wavy grain is typically grey-brown with darker streaks, although this can vary according to origin. It is reasonably tough, and is good for steam-bending. Italian walnut is considered to have the best colour and figure.

Common uses: Furniture, interior joinery, gun stocks, turnery, carving, veneer.

Workability: It can be worked well with hand and machine tools, and glues satisfactorily.

Finishing: It can be polished to a fine finish.

Average dried weight: 670kg/m^3 (42lb/ft^3).

AMERICAN WHITEWOOD

Liriodendron tulipifera

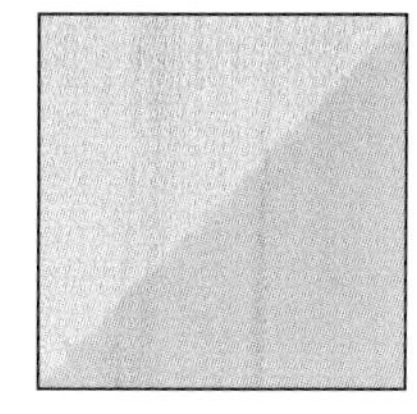

Other names: Canary whitewood (UK); yellow poplar, tulip poplar (USA); tulip tree.

Sources: Eastern USA, Canada.

Characteristics of the tree: It reaches a height of about 37m (125ft), and has an average diameter of 2m (6ft 6in).

Characteristics of the wood: The straight-grained, fine-textured wood is quite soft and lightweight; it is not durable, and should not be used in contact with the ground. The narrow sapwood is white; the heartwood ranges from pale olive-green to brown, streaked with blue.

Common uses: Light construction, interior joinery, toy-making, furniture, carving, plywood, veneer.

Workability: It can be worked easily with hand and machine tools, and glues well.

Finishing: It accepts stains, paint and varnish well, and can be polished to a good finish.

Average dried weight: 510kg/m^3 (31lb/ft^3).

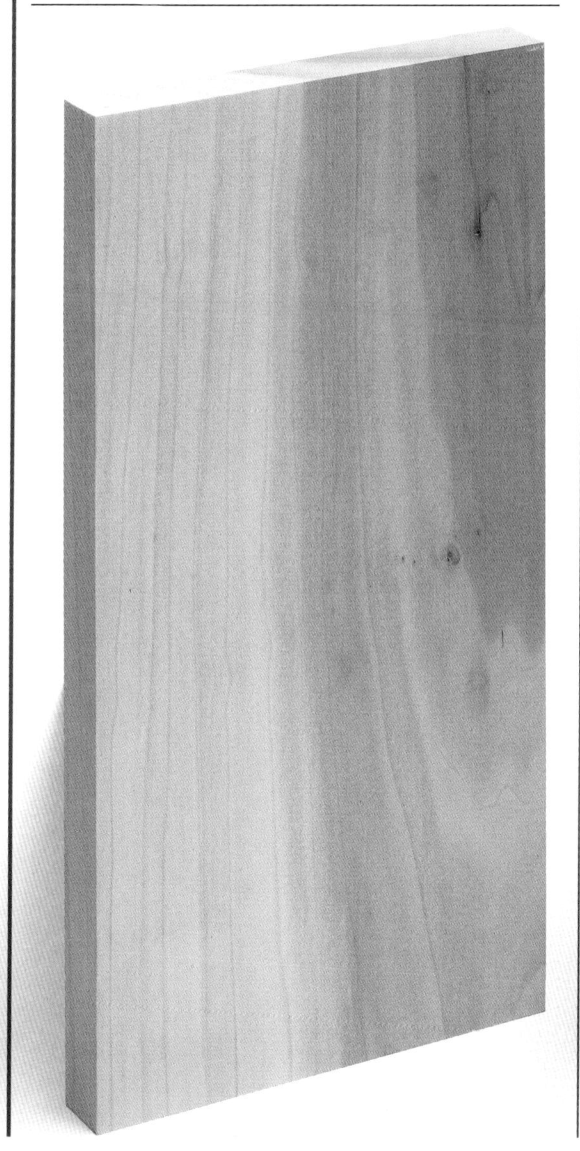

BALSA

Ochroma lagopus

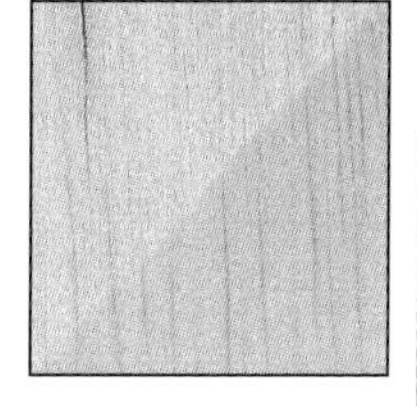

Other names: Guano (Puerto Rico, Honduras); topa (Peru); lanero (Cuba); tami (Bolivia); polak (Belize, Nicaragua).

Sources: South America, Central America, West Indies.

Characteristics of the tree: This fast-growing tree reaches a height of about 21m (70ft) and a diameter of about 600mm (2ft) in six to seven years, after which the growth rate declines. It reaches maturity in 12 to 15 years.

Characteristics of the wood: The open, straight-grained and lustrous wood is the lightest commercial hardwood, and is graded on its density: fast-grown wood is lighter in weight than the denser, harder wood produced by older, slower-growing trees. The colour is pale beige to pinkish.

Common uses: Insulation, buoyancy aids, model-making, packaging for delicate items.

Workability: If cutting edges are kept sharp to avoid crumbling or tearing, it can be worked and sanded easily with hand and machine tools; it glues well.

Finishing: It can be stained, painted and polished satisfactorily.

Average dried weight: 160kg/m^3 (10lb/ft^3).

PURPLEHEART

Peltogyne spp.

Other names: Amaranth (USA); pau roxo, amarante (Brazil); purperhart (Surinam); saka, koroboreli, sakavalli (Guyana).

Sources: Central America, South America.

Characteristics of the tree: This tall tree can reach a height of 50m (165ft), with a long, straight trunk about 1m (3ft 3in) in diameter.

Characteristics of the wood: The durable wood is strong and resilient, and has a uniform fine-to-medium texture; the grain is usually straight, but can be irregular. When first cut, the wood is a purple colour which darkens in time to rich brown through oxidation.

Common uses: Building construction, boatbuilding, furniture, turnery, flooring, veneer.

Workability: It can be worked well, although cutting edges must be kept sharp, as dull edges bring gummy resin to the surface. It is a good wood for turning, and glues well.

Finishing: It accepts stains well and can be wax-polished, but methylated-spirit (alcohol) -based polishes can affect the colour.

Average dried weight: 880kg/m^3 (55lb/ft^3).

● **ENDANGERED**

AFRORMOSIA

Pericopsis elata

Other names: Assemela (Ivory Coast, France); kokrodua (Ghana, Ivory Coast); ayin, egbi (Nigeria).

Sources: West Africa.

Characteristics of the tree: It is a long-trunked, relatively tall tree that reaches a height of about 45m (150ft) and a diameter of about 1m (3ft 3in).

Characteristics of the wood: The yellow-brown heartwood of this durable wood darkens to the colour of teak *(Tectona grandis)*; however, the straight-to-interlocked grain has a finer texture, and the wood is stronger and less oily than teak. In moist conditions, it can react with ferrous metals and develop black stains.

Common uses: Veneer, interior and exterior joinery and furniture, building construction, boatbuilding.

Workability: If care is taken with interlocking grain, it saws well and can be planed smooth. It glues well.

Finishing: It can be polished to a fine finish.

Average dried weight: 710kg/m^3 (44lb/ft^3).

● **ENDANGERED**

Afrormosia is listed in Appendix II in CITES (see page 14).

EUROPEAN PLANE

Platanus acerifolia

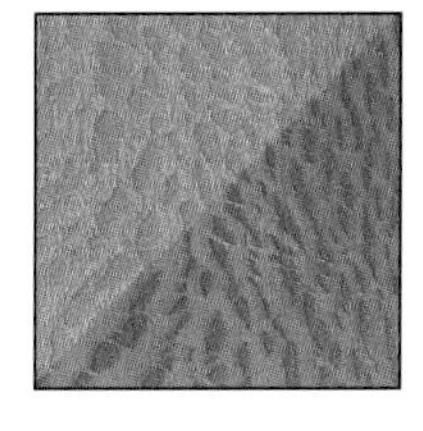

Other names: London, English, French plane etc., according to origin.

Sources: Europe.

Characteristics of the tree: Easily identified by its flaking, mottled bark, this tree is often found in cities, due to its tolerance of pollution. It grows to a height of about 30m (100ft), and produces a trunk about 1m (3ft 3in) in diameter.

Characteristics of the wood: The straight-grained, fine-to-medium-textured wood is perishable and not suitable for exterior use. The light reddish-brown heartwood has distinct darker rays; when quarter-sawn, these produce a fleck figure known as 'lacewood'. It is similar to, but darker than, American sycamore (see right), and is a good wood for steam-bending.

Common uses: Joinery, furniture, turnery, veneer.

Workability: It can be worked well with hand and machine tools, and glues well.

Finishing: It accepts stains and polishes satisfactorily.

Average dried weight: 640kg/m³ (40lb/ft³).

AMERICAN
SYCAMORE

Platanus occidentalis

Other names: Buttonwood (USA); American plane (UK).

Sources: USA.

Characteristics of the tree: This large tree grows up to 53m (175ft) in height and up to 6m (20ft) in diameter.

Characteristics of the wood: The fine, pale brown, even-textured wood is perishable and not suitable for exterior use; it is usually straight-grained, and distinct darker rays produce lacewood when quarter-sawn. It is botanically a plane tree, but is lighter in weight than European plane (see left).

Common uses: Joinery, doors, furniture, panelling, veneer.

Workability: The wood works well with hand and power tools; when planing it, keep cutters sharp. It can be glued well.

Finishing: It accepts stains and polishes satisfactorily.

Average dried weight: 560kg/m³ (35lb/ft³).

AMERICAN CHERRY

Prunus serotina

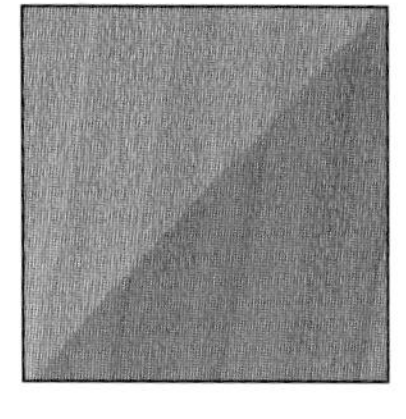

Other names: Cabinet cherry (Canada); black cherry (Canada, USA).

Sources: Canada, USA.

Characteristics of the tree: This moderate-size tree reaches a height of 21m (70ft), with a trunk about 500mm (1ft 8in) in diameter.

Characteristics of the wood: The durable wood has straight grain and a fine texture; it is hard and moderately strong, and can be steam-bent. The narrow sapwood is a pinkish colour, while the heartwood is reddish-brown to deep red, with brown flecks and some gum pockets.

Common uses: Furniture, pattern-making, joinery, turnery, musical instruments, tobacco pipes, veneers.

Workability: It can be worked well with hand and machine tools, and glues well.

Finishing: It accepts stain well and can be polished to a fine finish.

Average dried weight: 580kg/m^3 (36lb/ft^3).

AFRICAN PADAUK

Pterocarpus soyauxii

Other names: Camwood, barwood.

Sources: West Africa.

Characteristics of the tree: It grows to a height of 30m (100ft); the diameter of the trunk above the buttresses can reach 1m (3ft 3in).

Characteristics of the wood: The hard, heavy wood has straight-to-interlocked grain and a moderately coarse texture. The pale-beige sapwood can be 200mm (8in) thick, and the very durable heartwood is rich red to purple-brown, streaked with red.

Common uses: Interior joinery, furniture, flooring, turnery, handles; also used as a dyewood.

Workability: It can be worked well with hand tools, and can be machined to a fine finish on the surface; it glues well.

Finishing: It can be polished to a fine finish.

Average dried weight: 710kg/m^3 (44lb/ft^3).

● ENDANGERED

AMERICAN
WHITE OAK
Quercus alba

Other names: White oak (USA).

Sources: USA, Canada.

Characteristics of the tree: It can reach a height of 30m (100ft), and a diameter of about 1m (3ft 3in) in good growing conditions.

Characteristics of the wood: The straight-grained wood is similar in appearance to European oak *(Quercus robur)*, but is more varied in colour, which ranges from pale yellow-brown to pale brown, sometimes with a pinkish tint. It is reasonably durable for exterior use, and the texture is medium-coarse to coarse, again depending on the growing conditions. The wood has good steam-bending properties.

Common uses: Building construction, interior joinery, furniture, flooring, plywood, veneer.

Workability: It can be readily worked with hand and machine tools, and glues satisfactorily.

Finishing: It accepts stains well, and can be polished to a good finish.

Average dried weight: 770kg/m³ (48lb/ft³).

JAPANESE OAK
Quercus mongolica

Other names: Ohnara.

Sources: Japan.

Characteristics of the tree: It grows to a height of about 30m (100ft), and the straight trunk reaches a diameter of about 1m (3ft 3in).

Characteristics of the wood: The coarse texture of this straight-grained wood is milder than that of the European and American white oaks, due to its slower, more even rate of growth. The colour is light yellowish-brown throughout, and it is generally knot-free; it is a good wood for steam-bending, and the heartwood is moderately durable for exterior use.

Common uses: Interior and exterior joinery, boatbuilding, furniture, panelling, flooring, veneer.

Workability: Compared to other white oaks, it is easy to work well with hand and machine tools, and glues well.

Finishing: It accepts stain and can be polished very well.

Average dried weight: 670kg/m³ (41lb/ft³).

EUROPEAN OAK

Quercus robur/Q. petraea

Other names: English, French, Polish oak etc., according to origin.

Sources: Europe, Asia Minor, North Africa.

Characteristics of the tree: It can grow to above 30m (100ft) in height, and the trunk can be up to 2m (6ft 6in) in diameter.

Characteristics of the wood: The coarse-textured wood has straight grain, distinct growth rings, and broad rays that show an attractive figure when quarter-sawn. The sapwood is much paler than the pale yellowish-brown of the heartwood. It is a tough wood that is good for steam-bending. Although durable, it is acidic, and causes metals to corrode. Oaks grown in Central Europe tend to be lighter and less strong than those from Western Europe.

Common uses: Joinery and external woodwork, furniture, flooring, boatbuilding, carving, veneer.

Workability: If sharp cutting edges are maintained, it can be worked readily with hand and machine tools; it glues well.

Finishing: Liming, staining and fuming are all possible, and it can be polished to a good finish.

Average dried weight: 720kg/m^3 (45lb/ft^3).

AMERICAN RED OAK

Quercus rubra

Other names: Northern red oak.

Sources: Canada, USA.

Characteristics of the tree: Depending on the growing conditions, it can reach a height of 21m (70ft) and a diameter of 1m (3ft 3in).

Characteristics of the wood: The non-durable wood has straight grain and a coarse texture, though this can vary according to the rate of growth; northern wood is not as coarse as that from trees grown in southern states, which are faster-grown. The colour is a similar pale yellowish-brown to the white oaks, but with a pinkish-red hue. It is good for steam-bending.

Common uses: Interior joinery and flooring, furniture, decorative veneer, plywood.

Workability: It can be worked readily with hand and machine tools, and glues satisfactorily.

Finishing: It accepts stains well, and can be polished to a good finish.

Average dried weight: 790kg/m^3 (49lb/ft^3).

RED LAUAN

Shorea negrosensis

Other names: None.

Sources: Philippines.

Characteristics of the tree: This large tree can reach a height of 50m (165ft), with a long, straight trunk of about 2m (6ft 6in) diameter above the buttresses.

Characteristics of the wood: The wood is moderately durable, with interlocked grain and a relatively coarse texture; an attractive ribbon-grain figure is shown on quarter-sawn boards. The sapwood is a light creamy colour, while the heartwood is medium to dark red.

Common uses: Interior joinery, furniture, boatbuilding, veneer, boxes.

Workability: Although it can be worked easily with hand and machine tools, care must be taken that the surface of the wood does not tear when planing. It glues well.

Finishing: It accepts stains well, and can be varnished and polished to a good finish.

Average dried weight: 630kg/m³ (39lb/ft³).

● **ENDANGERED**

BRAZILIAN

MAHOGANY

Swietenia macrophylla

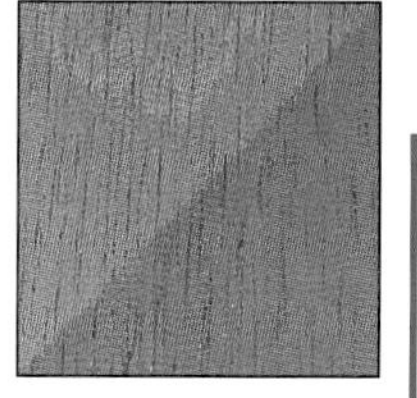

Other names: Honduras, Costa Rican, Peruvian mahogany etc.

Sources: Central America, Southern America.

Characteristics of the tree: This large tree can grow up to 45m (150ft) in height, and can reach about 2m (6ft 6in) in diameter above heavy trunk buttresses.

Characteristics of the wood: The naturally durable wood has a medium texture and grain that is either straight and even or interlocked. The white-yellow sapwood contrasts with the reddish-brown to deep red heartwood.

Common uses: Interior panelling, joinery, boat planking, furniture, pianos, carving, decorative veneer.

Workability: It can be worked well with hand and machine tools, if cutting edges are kept sharp; it glues well.

Finishing: It accepts stains very well, and can be polished to a fine finish when the grain is filled.

Average dried weight: 560kg/m³ (35lb/ft³).

Brazilian mahogany is listed in Appendix III in CITES (see page 14).

● **ENDANGERED**

TEAK

Tectona grandis

Other names: Kyun, sagwan, teku, teka.

Sources: Southern Asia, Southeast Asia, Africa, Caribbean.

Characteristics of the tree: It can reach 45m (150ft) in height, with a long, straight trunk about 1.5m (5ft) in diameter. The trunk can be fluted and buttressed.

Characteristics of the wood: The strong, very durable wood has a coarse, uneven texture with an oily feel; the grain can be straight or wavy. 'Burma' teak (from Myanmar) is a uniform golden-brown, while other areas produce a darker, more marked wood. It is a moderately good wood for steam-bending.

Common uses: Interior and exterior joinery, boatbuilding, exterior furniture, turnery, plywood, veneer.

Workability: It can be worked well with hand and machine tools, but quickly dulls cutting edges. Newly prepared surfaces glue well.

Finishing: It accepts stains, varnishes and polishes, and can be finished well with oil.

Average dried weight: 640kg/m³ (40lb/ft³).

● **ENDANGERED**

BASSWOOD

Tilia americana

Other names: American lime.

Sources: USA, Canada.

Characteristics of the tree: This medium-size tree averages a height of 20m (65ft) and a diameter of 600mm (2ft); the straight trunk is often free of branches for much of its length.

Characteristics of the wood: The straight-grained wood has a fine, even texture; it is not durable, and is lighter in weight than the related European lime *(Tilia vulgaris)*. There is little contrast between latewood and earlywood, and the soft, weak wood is cream-white when first cut, turning pale brown on exposure.

Common uses: Carving, turnery, joinery, pattern-making, piano keys, drawing boards, plywood.

Workability: It can be easily and cleanly worked with hand and machine tools, and can be brought to a fine surface finish. It glues well.

Finishing: It accepts stains well, and can be polished to a fine finish.

Average dried weight: 416kg/m³ (26lb/ft³).

LIME

Tilia vulgaris

Other names: Linden (Germany).

Sources: Europe.

Characteristics of the tree: It can reach a height of more than 30m (100ft), and the clear trunk is about 1.2m (4ft) in diameter.

Characteristics of the wood: The straight-grained wood has a fine, uniform texture; although soft, it is strong and resists splitting, making it particularly good for carving and turning. It is perishable, but can be treated with preservative. The overall colour is white to pale yellow, but darkens to light brown with exposure; there is no distinction between sapwood and heartwood.

Common uses: Carving, turnery, toy-making, hat blocks, broom handles, harps, piano soundboards and keys.

Workability: It is an easy wood to work with hand and machine tools, as long as cutting edges are kept sharp; it glues well.

Finishing: It accepts stains well, and can be polished to a fine finish.

Average dried weight: 560kg/m^3 (35lb/ft^3).

OBECHE

Triplochiton scleroxylon

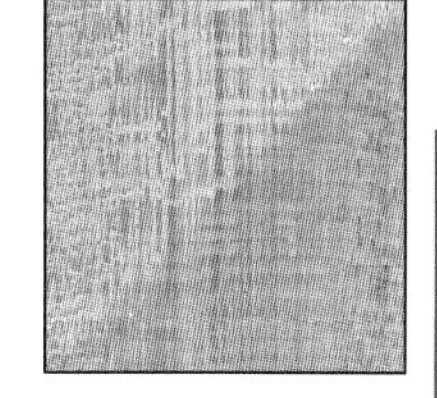

Other names: Ayous (Cameroon); wawa (Ghana); obechi, arere (Nigeria); samba, wawa (Ivory Coast).

Sources: West Africa.

Characteristics of the tree: This large tree can be above 45m (150ft) in height, and the trunk is about 1.5m (5ft) in diameter above heavy buttresses.

Characteristics of the wood: The fine, even-textured wood is lightweight and not durable; the grain can be straight or interlocked. There is little contrast between the sapwood and heartwood; both are cream-white to pale yellow in colour.

Common uses: Interior joinery, furniture, drawer linings, model-making, plywood.

Workability: If cutting edges are kept sharp, the soft wood is easy to work with hand and machine tools. It glues well.

Finishes: It accepts stains and polishes well.

Average dried weight: 390kg/m^3 (24lb/ft^3).

● **ENDANGERED**

AMERICAN WHITE ELM

Ulmus americana

Other names: Water elm, swamp elm, soft elm (USA); orhamwood (Canada).

Sources: Canada, USA.

Characteristics of the tree: This medium-to-large tree usually reaches a height of 27m (90ft), with a trunk 500mm (1ft 8in) in diameter, but good growing conditions can produce larger trees.

Characteristics of the wood: The coarse-textured wood is not durable; it is strong, tougher than the European elms (see right) and, like them, good for steam-bending. The grain is usually straight, but can be interlocked, and the heartwood is pale reddish-brown.

Common uses: Boatbuilding, agricultural implements, cooperage, furniture, veneer.

Workability: Sharp cutting edges enable it to be readily worked with hand and machine tools, and it glues satisfactorily.

Finishing: It accepts stains, and polishes satisfactorily.

Average dried weight: 580kg/m^3 (36lb/ft^3).

DUTCH AND ENGLISH ELM

Ulmus hollandica/U. procera

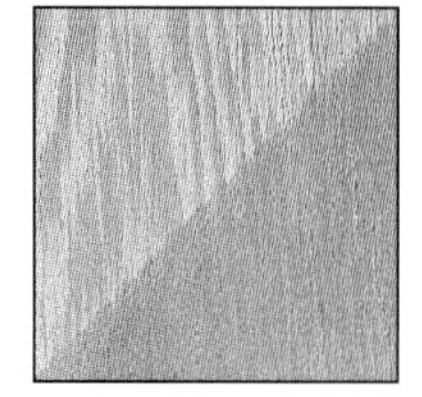

Other names: *English elm:* Red elm. *Dutch elm:* Cork bark elm.

Sources: Europe.

Characteristics of the tree: This relatively large tree can reach 45m (150ft) in height and up to 2.5m (8ft) in diameter, but elms are usually cut when they reach a diameter of about 1m (3ft 3in).

Characteristics of the wood: The coarse-textured wood has beige-brown heartwood and distinct irregular growth rings, with an attractive figure when plain-sawn. It is not durable. The Dutch elm is tougher than the English, with more even growth and straighter grain, making it better for steam-bending. Dutch elm disease has led to short supplies of the wood.

Common uses: Cabinetmaking, Windsor-chair seats and backs, boat-building, turnery, veneer.

Workability: Irregular-grain wood can be difficult to work, especially when planing, but it can be brought to a smooth surface finish. It glues well.

Finishing: It accepts stains and polishes well, and is particularly suited to a wax finish.

Average dried weight: 560kg/m^3 (35lb/ft^3).

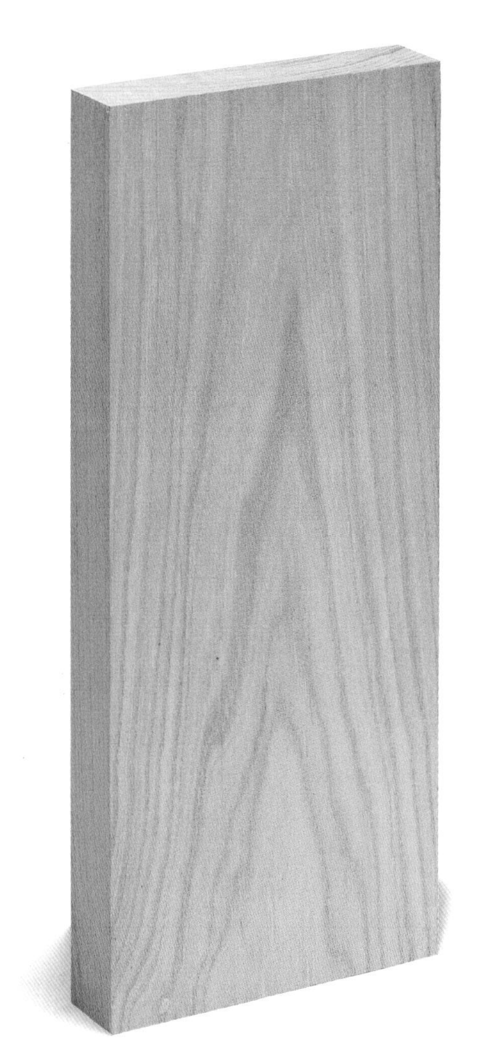

WOOD VENEERS

CHAPTER 3 Veneers are very thin sheets of wood, known as leaves, which are cut or sliced from a log for constructional or decorative purposes. Whether they are selected for their natural colour and figure or worked into formal patterns, veneers bring a unique quality to furniture and woodware.

VENEER PRODUCTION

With the widespread use of stable man-made boards for groundwork, and the development of modern adhesives, today's veneered products are superior to solid wood for certain applications. Highly sophisticated production techniques are used to satisfy the growing demand for veneer.

Choosing logs

Every stage in the manufacture of veneer requires specialist knowledge. The process starts with the log buyer, who must have the skill and experience to assess the condition and commercial viability for veneer within a log, basing this solely on an external examination. By looking at the end of the log, he or she must determine the quality of the wood, the potential figure of the veneer, the colour and the ratio of sapwood to heartwood. Other factors, such as the presence and extent of staining, and weaknesses or defects in the form of shakes, ingrown bark, excessive knots or resin ducts, will affect the value or suitability of the log, and must also be taken into consideration.

Much of this information will be revealed by the first cut through the length of the log; however, the log must be purchased before this first cut can be made.

Decorative-veneer wall panelling and moulded-plywood seating

KNIFE CHECKS

Veneer-slicing machines cut like giant planes; it is vital that the shaving is produced to fine tolerances and with a clean cut. The quality of the cut is controlled by the pressure bar and knife setting.

Fine cracks known as knife checks can occur on the back face of the veneer, particularly when it has been rotary-cut (see page 86).

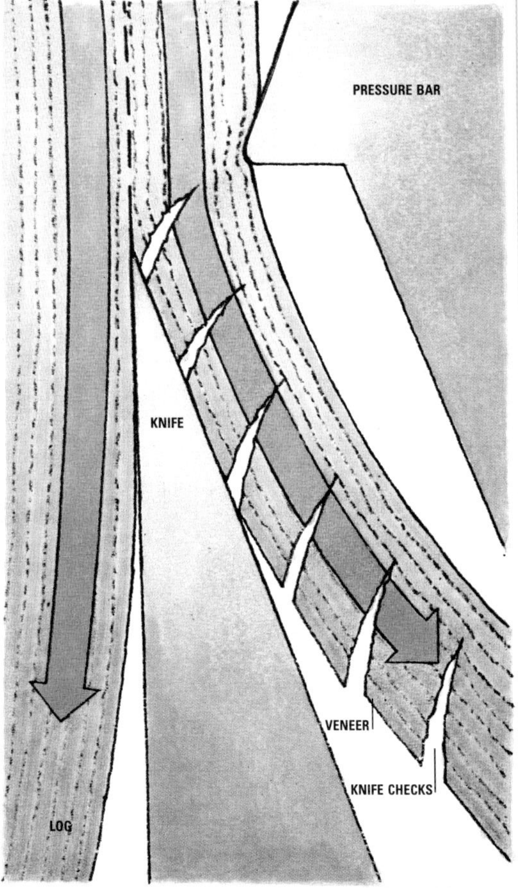

Open and closed faces

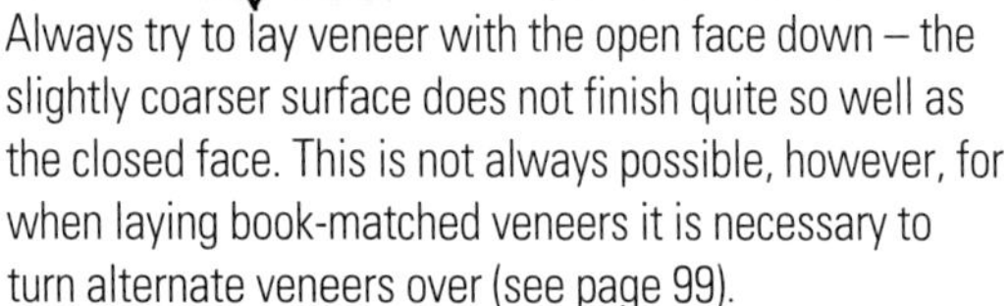

The back face of the veneer is called the open, or loose, face, and the other the closed, or tight, face. The faces can be identified by flexing the veneer along the grain; it will bend more when the open face is convex.

Always try to lay veneer with the open face down – the slightly coarser surface does not finish quite so well as the closed face. This is not always possible, however, for when laying book-matched veneers it is necessary to turn alternate veneers over (see page 99).

Treating logs

Logs are softened, either by immersion in hot water or by being steamed, before they are converted into veneer. Depending on the cutting method, the whole log may be treated or may first be cut into flitches by a huge bandsaw.

The time taken for this softening is controlled by the type and hardness of the wood and the thickness of the veneer to be cut. The process can take days or weeks.

Some pale woods, such as maple and sycamore, are not pre-treated because the softening process would discolour the veneer.

Cutting veneer

Another skilled production expert is the veneer cutter, who decides the best way to convert the log so that it produces the maximum number of high-quality leaves.

Most veneer logs are cut from the main stem of the tree between the root butt and the first branch. The bark is removed and the log is checked for foreign matter, such as nails or wire.

As decorative veneers are cut, they are taken from the slicer and stacked in sequence. This stack, or set, then passes through a machine-drying process before being graded.

Although most species are clipped on a guillotine to trim them into regular shapes and sizes, others, such as yew or burr veneer, are kept as when cut from the log.

Grading decorative veneer

Veneers are graded according to their size and quality, checked for natural or milling defects, thickness, colour and type of figure, among other features, and are priced accordingly. The veneers from one particular log can be of various values; the better or wider ones are graded as face quality, and have a greater value than narrower or poorer backing quality, or balancing, veneer.

The veneers are kept in multiples of four for matching purposes, and are bound into bundles of 16, 24, 28 or 32 leaves. The bundles are restacked in consecutive order, and the reassembled log is stored in a cool warehouse, ready for sale.

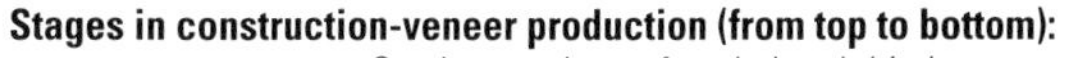

Stages in construction-veneer production (from top to bottom):
Continuous sheet of peeled-and-dried veneer; patching cut-and-dried veneer (this is often an automated process); veneer being glued and stacked before pressing into plywood.

CUTTING METHODS

The three basic methods for cutting veneer are saw cutting, rotary cutting, and flat slicing; there are also variations on the latter two techniques.

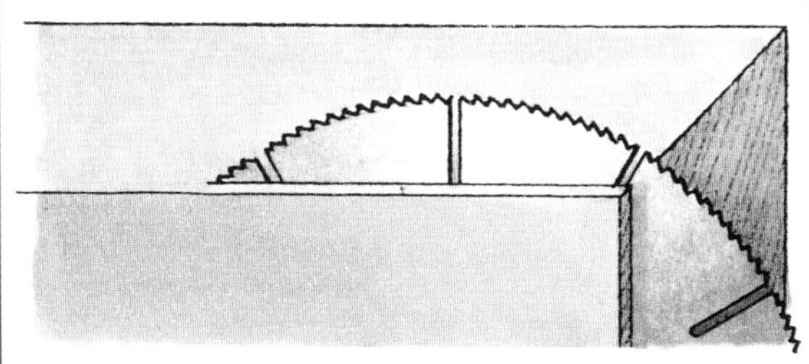

Saw cutting

Until the development of veneer-slicing machines, all veneers were cut using saws, first by hand and then by powered circular saws. The veneers thus cut were relatively thick, around 3mm (⅛in).

Sawn veneers are still produced, using huge circular saws, but only for very hard woods, such as lignum vitae, irregular-grained woods such as curls, or where sawing proves the most economical method, despite generally being a wasteful process. These veneers are usually about 1 to 1.2mm (1⁄25 to 1⁄21in) thick.

A workshop bandsaw or table saw can be used to produce strips of veneer for laminating purposes, particularly if this proves more economical or provides better matched material than is commercially available.

Flat-slicing hardwood veneer

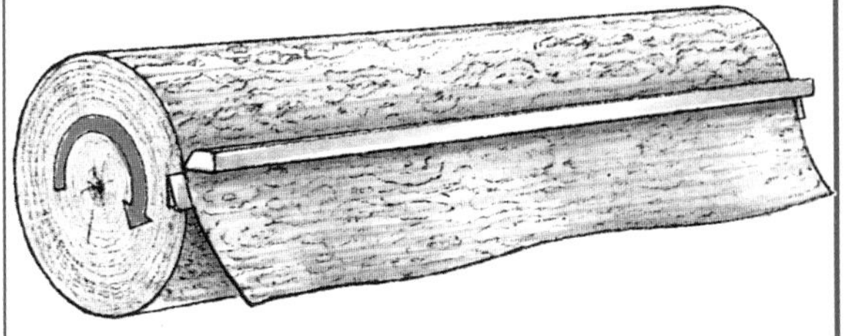

Rotary cutting

Although mostly used for making constructional veneers in softwood and some hardwoods, the rotary-cutting method is also used to produce decorative veneers, such as bird's-eye maple.

A complete log is set in a huge lathe which peels off a continuous sheet of veneer. The log is rotated against a pressure bar and knife which run the full length of the machine; the knife is set just below the bar and forward of it by the thickness of the veneer. The settings of the bar and knife are critical if 'checks' (see page 88) are to be prevented. The knife automatically advances by the thickness of the veneer for each revolution of the log.

Veneer thus produced can be identified by a distinctive watery patterned figure where the continuous tangential cut has sliced through the growth rings.

Rotary cutting is particularly suitable for the manufacture of man-made boards, as the veneer can be cut to any width.

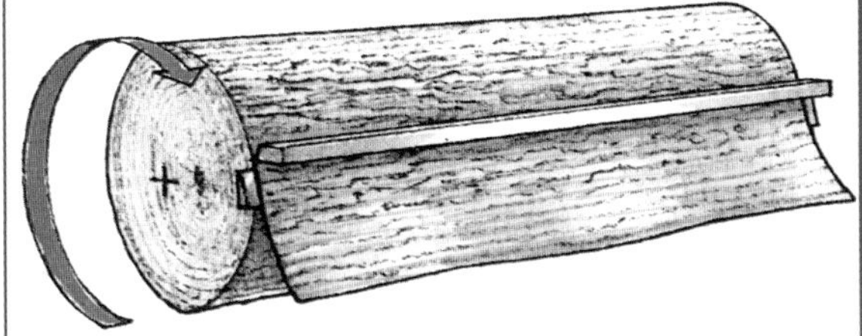

Off-centre cutting

By offsetting a log in rotary-lathe chucks to produce an eccentric cutting action, the rotary lathe can also produce wide, decorative veneers with sapwood on each edge; this results in a figure something like typical flat-sliced, crown-cut veneers.

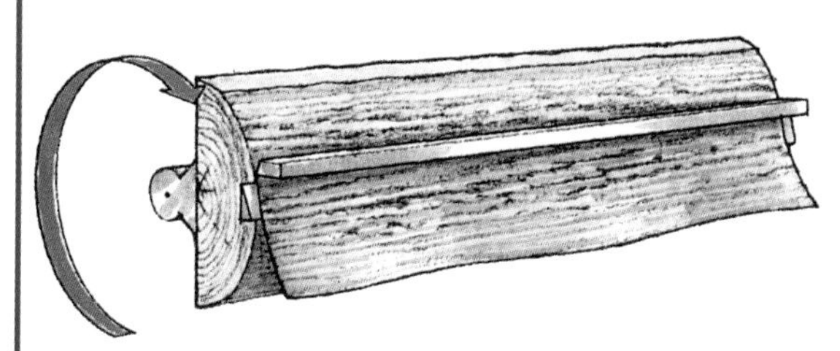

Half-round cutting

A mounting held between the lathe centres to which a full or half-round log is fixed is called a 'stay-log'. Veneers cut on a stay-log are sliced at a shallower angle than those taken from an eccentrically mounted log, but are not so wide. The figure produced is close to that of flat-sliced, crown-cut veneer.

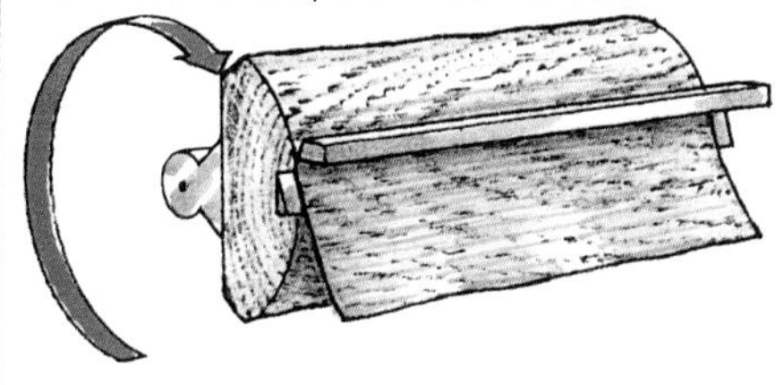

Back cutting

'Back cutting' is when half-round logs are mounted on a stay-log with the heartwood facing outwards, and is used for cutting decoratively figured butts and curls.

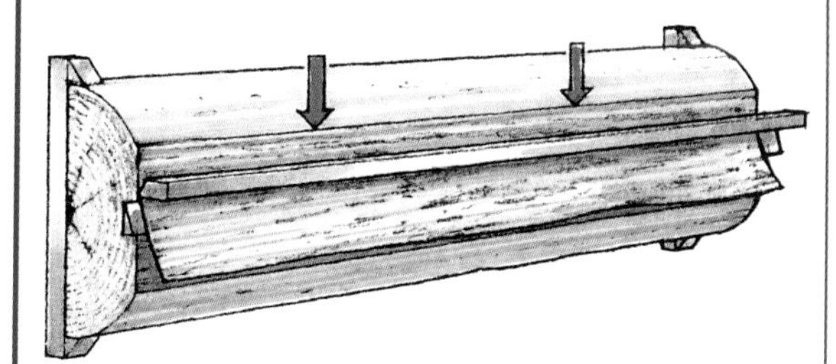

Flat slicing

The flat-slicing method is used to produce decorative hardwood veneers. A log is first cut in two through its length and the grain is assessed for figure. It may then be further cut into flitches, according to the type of figure required. The character of the figure depends on the way the log is cut and mounted for slicing, and the width of flat-sliced veneer is determined by the size of the flitch.

A half log or quartered flitch is mounted on a vertically sliding frame. A pressure bar and knife are set horizontally in front of the wood, removing a slice of veneer with every downstroke of the frame. Depending on the type of machine, either the knife or the flitch is advanced by the required thickness after each cut.

A flat-sliced half-round log produces the crown-cut veneers commonly used in cabinet work; these have the same figure as tangentially cut flat-sawn boards.

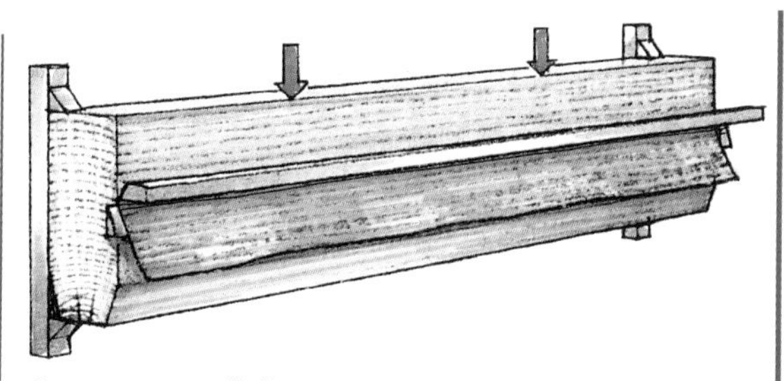

Quarter-cut slicing

Woods that display striking and attractive figure when radially cut are converted into quarter-cut or near-quarter-cut flitches. These are mounted so that the rays of the wood follow the direction of the cut as far as possible, to produce the maximum number of radially cut veneers.

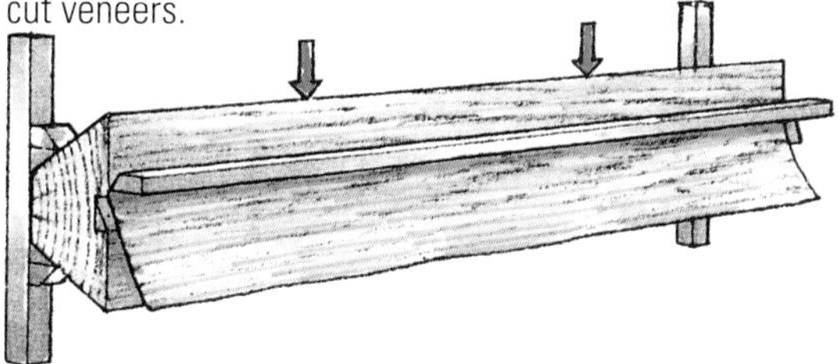

Tangential flat slicing

Quartered flitches can also be mounted to produce tangentially cut flat-sliced veneers. Although these are not as wide as crown-cut veneers cut from half-round logs, they can display attractive figure.

TYPES OF VENEER

Veneer figure derives both from the wood's natural features and from where and how it is cut. The description can refer to the method of cutting, such as 'crown-cut', or to the part of the tree from which veneer is cut, as in 'burr' veneer. Most decorative sliced veneer is cut about 0.6mm (1/42in) thick; 'construction' veneer, from 1.5 to 6mm (1/16 to 1/4in) thick, is also produced.

Buying veneer

Each veneer is unique and is unlikely to be matched from other bundles, so a generous allowance for wastage should be made when the area needed for a project is calculated.

Leaves are traditionally priced by the square foot, although the thickness may be given in metric measurements; some merchants supply pre-cut lengths at a set price per piece.

Veneer is almost always kept in consecutive order for matching purposes, so leaves and bundles are taken from the top of the stack. For the most part, suppliers do not pull out selected leaves, as doing so reduces the value of the veneer flitch.

Mail-order veneer

Orders of full veneers supplied through the mail are usually rolled for despatch. Smaller pieces, such as burrs or curls, may be flat-packed; if they are sent with a package of rolled veneer, they may be dampened to allow them to bend without breaking.

A rolled package should be opened carefully, to prevent it springing open and damaging the fragile veneer inside. End splits, particularly on light-coloured woods, must be repaired promptly with gummed paper tape, to prevent dirt getting into the split.

Veneer that is still curled after being unpacked can be dampened with steam from a kettle or passed through a tray of water, and can then be pressed flat between sheets of chipboard. Damp veneer left between boards can develop mildew.

Because wood is light-sensitive and can lighten or darken, according to species, veneers should be stored flat and protected from dust and strong light.

Inspecting veneer

Veneer should be inspected thoroughly for faults such as rough or open grain, splits or knife checks through the veneer, knife marks from a chipped blade, worm holes and hard inclusions in the pores.

BURR OR BURL VENEER

Burrs or burls are abnormal growths on tree trunks; the fragile veneers cut from them display an attractive pattern of tightly packed bud formations which appear as rings and dots. Burr veneer is highly prized for furniture, turnery and small woodware, and is, as a result, relatively expensive. It is supplied in irregular shapes in various sizes, from 150mm to 1m (6in to 3ft 3in) long, and from 100 to 450mm (4in to 1ft 6in) wide.

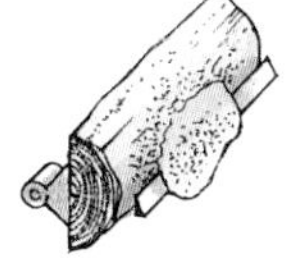

Top to bottom:
Elm burr; thuya burr; ash burr.

Part of tree:
Burr.

Method of cutting:
Rotary-cut or flat-sliced.

BUTT VENEER

Butt veneers are cut from stumps, or butts; half-round cutting on a rotary lathe produces highly figured veneers caused by distorted grain. They are fragile veneers, and may have holes where small pieces have become detached. To repair very small holes, matching filler can be applied after the veneer is laid.

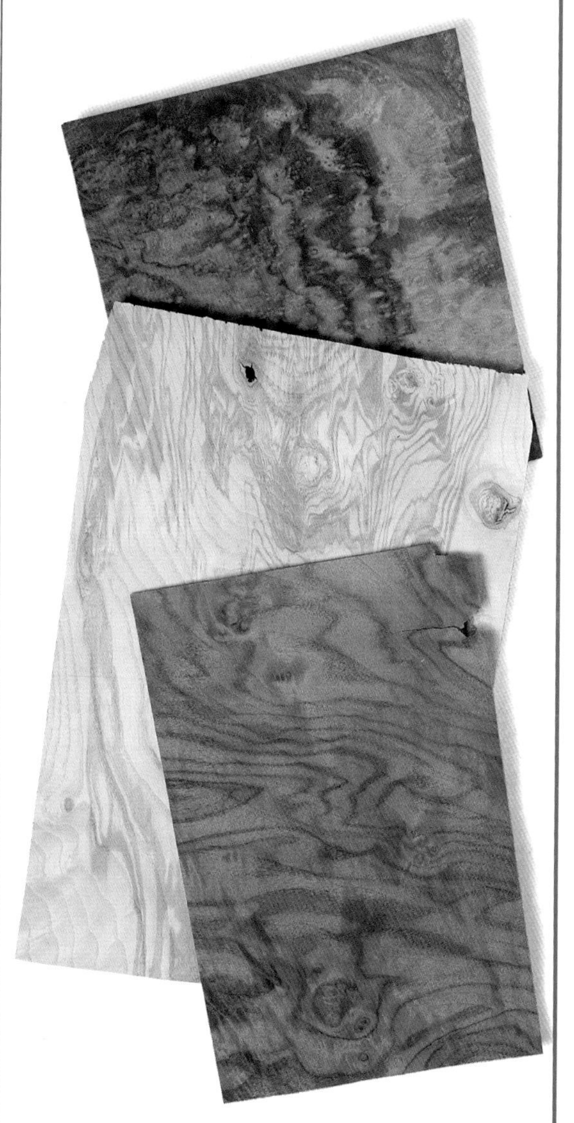

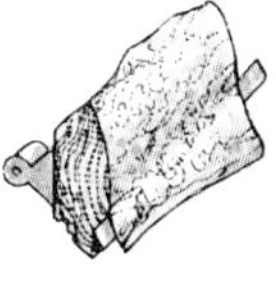

Top to bottom:
American walnut burry-butt veneer; ash butt veneer; American walnut butt veneer.

Part of tree:
Stump.

Method of cutting:
Rotary back-cut.

CROWN-CUT VENEER

Veneer that is tangentially cut and flat-sliced is called crown-cut, and has an attractive figure of bold sweeping curves and ovals along the centre of the leaf, with striped grain nearer the edges. It is produced in lengths of 2.4m (8ft) or more; widths range from about 225 to 600mm (9in to 2ft), depending on the species used. Crown-cut veneers are used for furniture-making and interior wall panelling.

Left to right:
Crown-cut kingwood; crown-cut Brazilian rosewood; crown-cut ash; crown-cut American walnut.

Part of tree:
Trunk.

Method of cutting:
Flat-sliced.

CURLY-FIGURED VENEER

Curly-figured veneers are produced by wavy-grained woods, and have bands of light and dark grain running across the width of the leaf. 'Fiddleback' sycamore and ripple ash are typical examples of this veneer; the former gets its name from its use in making violin backs. Curly-figured veneer is employed to give a distinctive horizontal decorative effect, for instance on cabinet doors and panels.

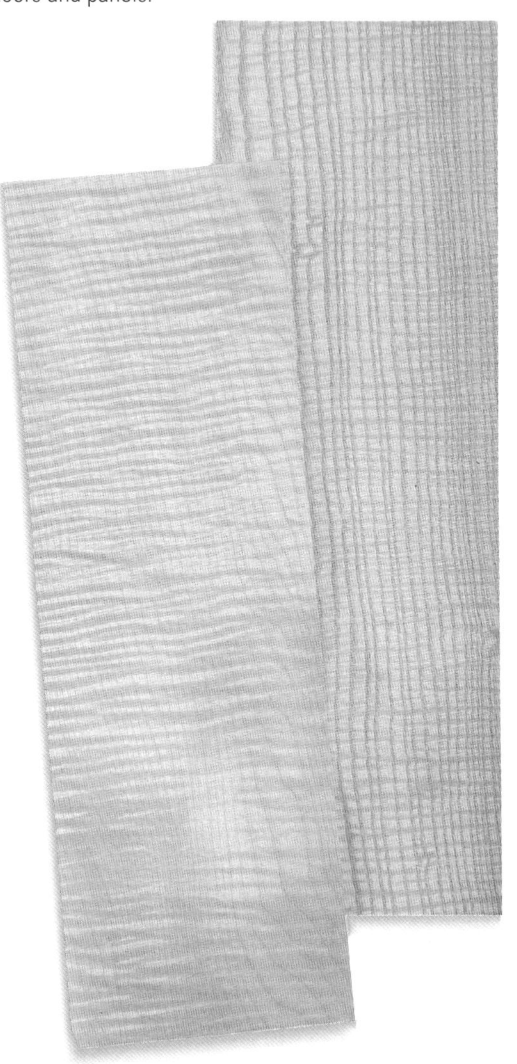

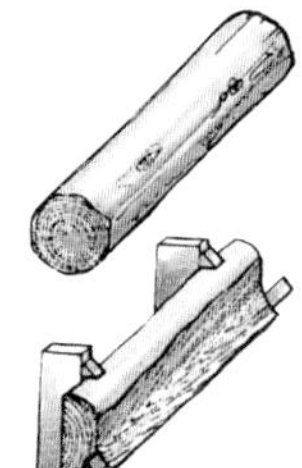

Left to right:
Fiddleback sycamore; ripple ash.

Part of tree:
Trunk.

Method of cutting:
Flat-sliced.

CURL VENEER

When the 'crotch' or fork of a tree, where the trunk divides, is cut perpendicularly, the attractive figure of curl veneer is revealed. The distorted diverging grain of the wood produces a lustrous upward-sweeping plume pattern known as 'feather figure'. Its striking features are often used on panelled cabinet doors. Curl veneer is available in lengths from 300mm to 1m (1ft to 3ft 3in) and widths from 200 to 450mm (8in to 1ft 6in).

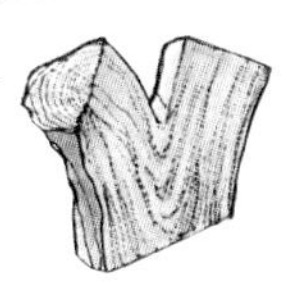

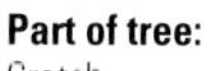

Top to bottom:
Mahogany curl; European walnut curl.

Part of tree:
Crotch.

Method of cutting:
Rotary back-cut.

FREAK-FIGURED VENEER

Veneers displaying various unusual patterns are rotary-cut from hardwood logs with irregular growth. Bird's-eye maple and masur birch are well-known types of such freak-figured veneers. The distinctive brown marks of masur birch, for example, are caused by wood-boring larvae that attack the cambium layer of the growing tree. Woods with irregular grain also produce veneers with 'blistered figure' and 'quilted figure'.

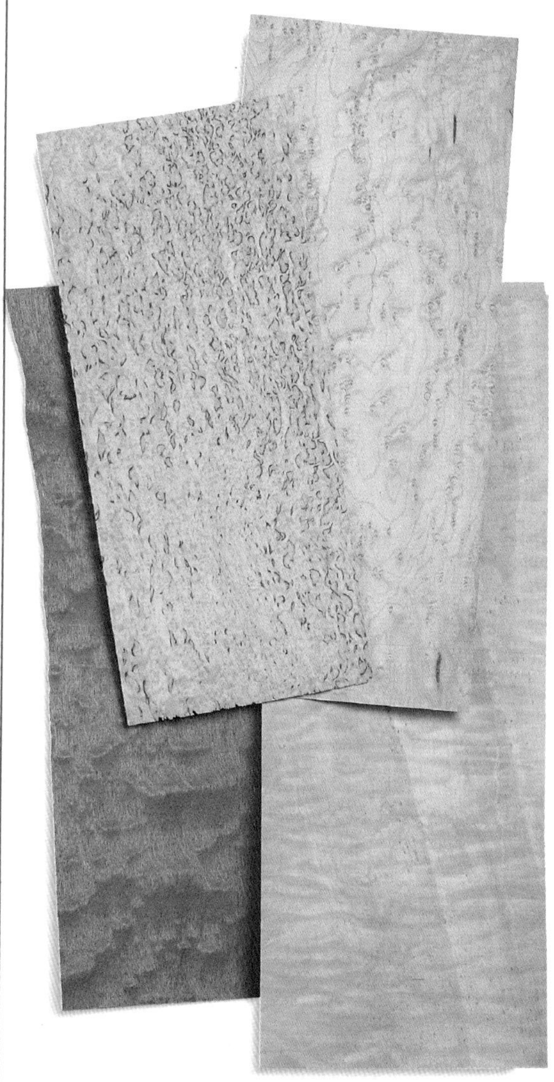

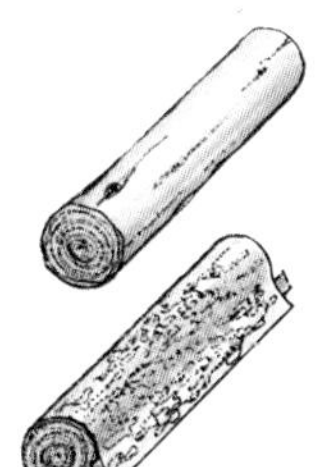

Left to right:
Quilted makoré; masur birch; bird's eye maple; quilted willow.

Part of tree:
Trunk.

Method of cutting:
Rotary-cut (peeled).

RAY-FIGURED VENEER

When quarter-cut, woods that have a pronounced ray-cell structure, such as oak and plane, display striking figure. Quarter-cut plane veneer, with speckled or fine wavy-grain figure, is known as lacewood; distinctive wide ray cells in oak produce ray-fleck, or splash, figure, long in demand for furniture-making and panelling. Ray-figured veneer is available in lengths of up to 2.4m (8ft) and in widths of between 150 and 350mm (6in and 1ft 2in), depending on the species used.

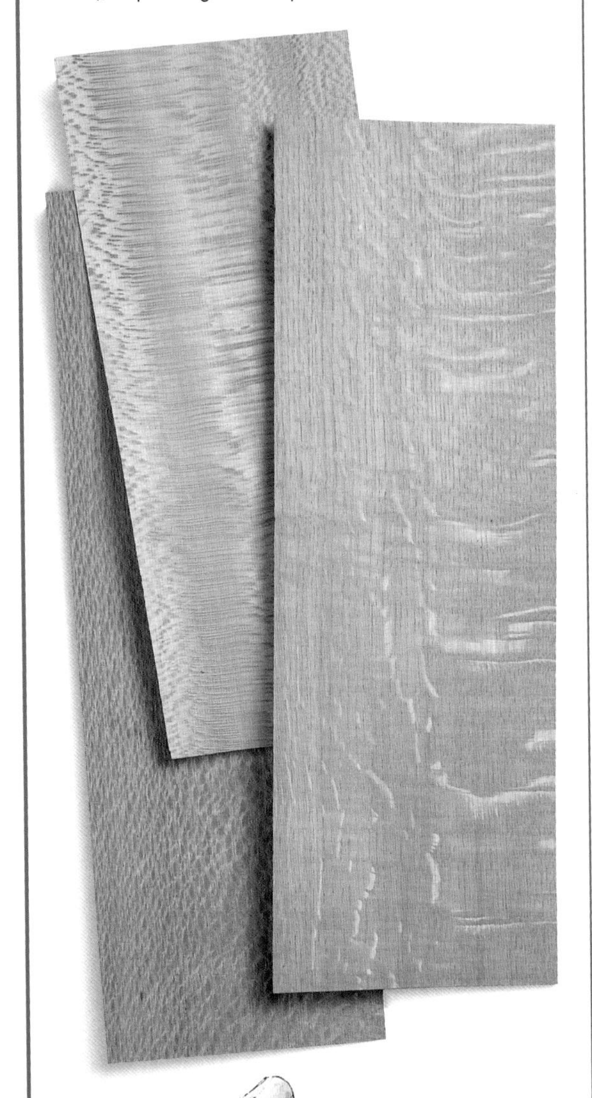

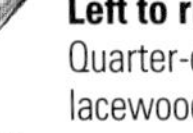

Left to right:
Quarter-cut silky oak; quarter-cut lacewood; quarter-cut ray-fleck oak.

Part of tree:
Trunk.

Method of cutting:
Quarter-cut flat-sliced.

STRIPED OR RIBBON VENEER

Where the radial cut is taken across the width of the growth rings, quarter-cut veneers usually display a striped, or ribbon, figure. Striped quarter-cut veneers are produced in lengths of 2.4m (8ft) or more, and widths of 150 to 225mm (6 to 9in). On woods which grow with interlocked reverse-spiral grain, the stripes appear to change from light to dark, depending on the degree of reflectivity of the cells (end-on cells absorb light), and the angle from which they are viewed.

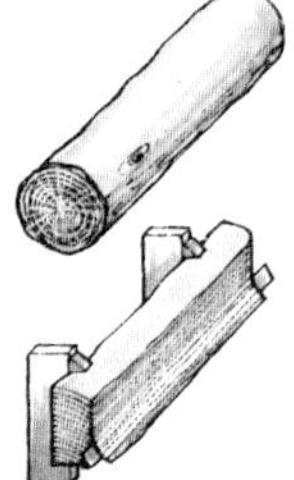

Left to right:
Stripe-figured zebrano; ribbon-figured sapele; ribbon-figured ayan.

Part of tree:
Trunk.

Method of cutting:
Quarter-cut flat-sliced.

COLOURED VENEER

Specialist veneer suppliers sell artificially coloured veneers made from light-coloured woods such as sycamore and poplar. Harewood is sycamore that has been chemically treated, turning it silver-grey to dark grey in colour. Other colours are produced by pressure-treating dyed veneers to achieve maximum penetration.

Clockwise from top left:
Turquoise-dyed veneer; blue-dyed veneer; chemically coloured harewood; green-dyed veneer.

Method of colouring:
Harewood (sycamore) is immersed in a ferrous-sulphate solution. Commercially colour-dyed woods are processed in an autoclave.

RECONSTRUCTED VENEER

Spectacular effects of colour, grain and pattern can be achieved using a computerized process that involves scanning, dyeing, high-pressure gluing and pressing veneer into a solid block and then re-slicing it into conventional and decorative-veneer patterns. Single leaves are produced up to 700mm (2ft 4in) wide and between 2500 and 3400mm (8ft 4in and 11ft 4in) long, and range from 0.3 to 3mm (1/84 to 1/8in) thick.

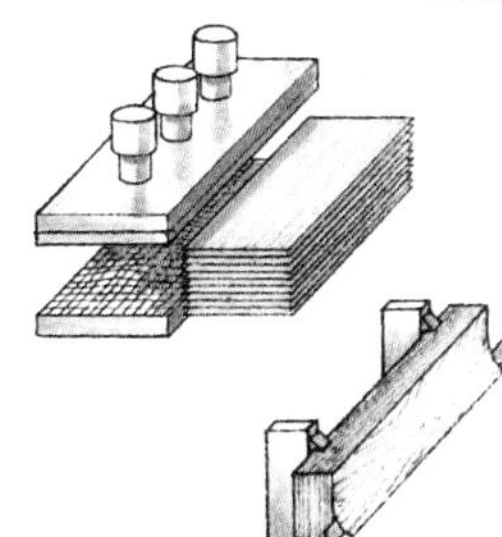

Top to bottom:
Geometric-stripe pattern; tessellated geometric pattern; freak-figured pattern; crown-cut pattern.

Method of manufacture:
Pressed veneers.

Method of cutting:
Flat-sliced.

LINES AND BANDINGS

Selected woods are used to produce decorative inlay stringing lines and bandings, which are usually sold in 1m (3ft 3in) lengths. Because the woods and sizes can vary between batches, it is always worth buying more than will be used for a particular project.

Stringing lines
The fine strips of wood used to divide areas of veneer are called stringings. They are made in flat and square sections, with the latter used for inlaying edges, and provide light or dark lines between different kinds of veneer or where the grain direction changes. Although boxwood and ebony were traditionally used for making stringings, black-dyed wood is now more common.

Bandings
Side-grain sections of coloured woods are glued together and sliced to produce decorative banding strips about 1mm (1/25in) thick and available in various widths. Used for ornamental borders, the strips are ready-edged, with boxwood or black stringings. Cross-bandings are strips of veneer cut across the grain, which are used for bordered panels. They can be made by either cutting from the veneer used for the panel, or from another suitable straight-grain veneer.

Stringing lines

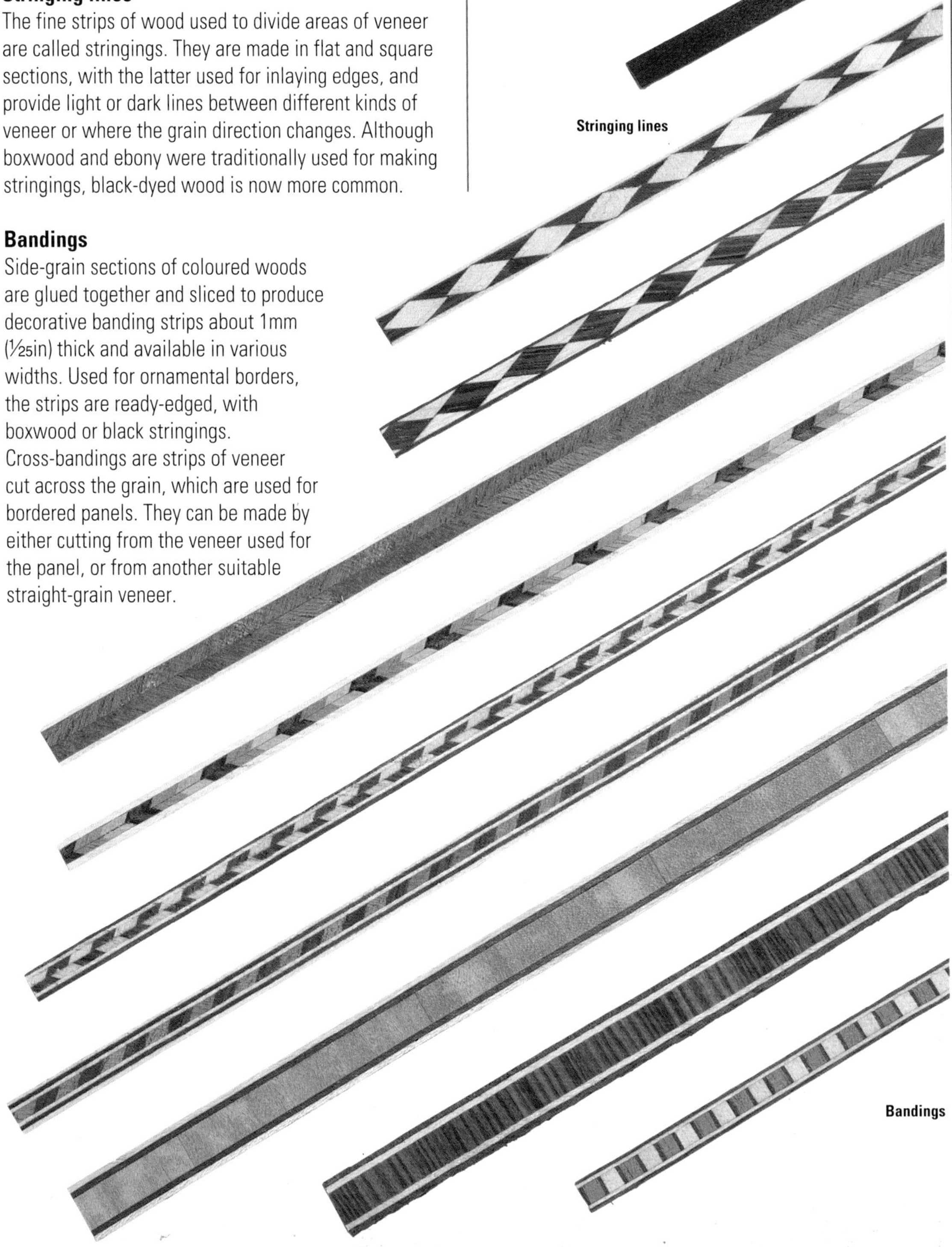

Bandings

VENEERING TOOLS

A basic woodworking tool kit will contain some tools used in veneer work, including measuring and marking tools, a fret saw, bench, block and shoulder planes, chisels, scrapers and sanding equipment. Most of the specialized tools for veneer work shown here are available from good tool stores or from specialist suppliers. You will need to make some equipment, such as a shooting board.

A selection of veneering tools, anticlockwise from top

Glue pot
A double or jacketed glue pot is used to prepare the hot animal glue used in traditional veneer laying. The outer pot holds water, which is heated to keep the glue in the inner pot at working temperature and prevent it burning. Older glue pots are made of cast iron; most modern versions are made from aluminium, and are heated by a source such as a gas or electric ring. Thermostatically controlled electric glue pots are available, but are more expensive.

Toothing plane
A toothing plane is used to 'key' the surface of the groundwork ready for gluing. The blade is set almost vertical; the front face is finely grooved, and a bevel ground on the back face produces a series of teeth across the blade, similar to a fine saw. The blade is sharpened by honing the bevel.

Veneer punches
There are eight sizes of veneer punches used to repair defects in veneer. An irregular-shape cutter makes a hole in the defective veneer and an identical patch from matching veneer. The cut patch is pushed from the tool by a sprung ejector.

Veneer pins
Fine short pins with large plastic heads are used to temporarily hold veneers while the joints are being taped.

Veneer saw
A veneer saw is used with a straightedge to cut any thickness of veneer. It produces a square-edged cut for accurately butt-jointing matched veneers. The double-edged blade is reversible, and the fine teeth have no set.

Knives
Surgical scalpels or craft knives fitted with pointed blades cut intricate shapes. The blades are ground on both sides to produce a 'V' cut. If the edge of the veneer must be cut square, the knife is held at an angle away from the straightedge. A stiffer, chisel-ground knife is used for straight cutting, particularly if greater pressure is required.

Rules and straightedges
A metal rule can double as a straightedge for small work. Pressed-steel 'safety rulers' have edges to grip the work and prevent it from slipping when used as a cutting guide. A groove folded along the centre keeps fingers safe. These rules are graduated in centimetres and inches. A steel straightedge is used for cutting longer veneers.

Veneer tape
Gummed paper tape is used to hold pieces of veneer together and to prevent newly laid veneer joints from opening up due to shrinkage. It is removed by wetting and scraping after the glue has set.

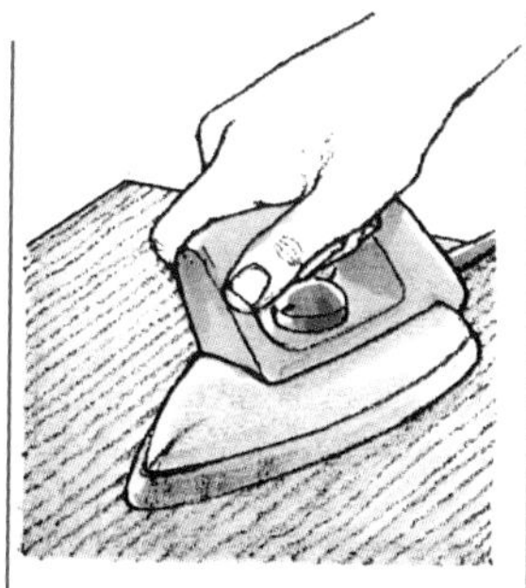

Electric iron
An old domestic electric iron can be used to soften animal glue applied to groundwork and veneer for hammer-veneering.

Veneer hammers
Used for hand-laying veneers, wooden-headed veneer hammers have a rounded brass blade mounted into a hardwood head fitted with a handle. Metal-headed hammers look more conventional, but the head is designed for pressing blisters.

Trimming tool
Veneer-trimming tools remove surplus veneer around the edges of a panel. On a typical tool, the short chisel-ground blade is adjustable, and cleanly cuts with or across the grain of the veneer.

Cutting mat
Proprietary cutting mats are made of self-sealing rubber compound. Knife points can cut into the surface without it becoming permanently scored or dulling the blade. Fine-surfaced man-made boards, such as hardboard, are good alternatives.

GROUNDWORK

The 'ground' or 'groundwork' – the backing material to which veneer is glued – is as important as the choice of veneer itself. Whether it is flat or curved, the ground should provide a smooth, even surface that is free from marks and dust. Thin veneer will not disguise or mask any surface defects or unevenness; in fact they will 'telegraph' or show through it, particularly when the finished work is polished.

Selecting a ground

Despite the tradition of using pine or mahogany as a base for veneered furniture, solid wood is not the ideal ground for veneering; it 'moves' due to changes in humidity, and requires careful preparation, particularly when making up wide panels. It has largely been superseded by man-made boards (see pages 106–14), which are stable and easy to prepare; they are made with flat, sanded surfaces, and can be obtained in large sheets.

Veneering solid wood

To ensure that veneer and solid wood 'move' together and are therefore more stable, veneer must be laid in line with the grain of the solid material. Quarter-cut boards are the most stable, as there is only slight shrinkage across the width and thickness. Laying veneer on both sides of the board is best, as this helps maintain an even balance.

However, if only one side of a tangentially cut board is to be veneered, this should always be the 'heart' side. As the glue dries, the veneer helps pull the board flat and resists any tendency to 'cup'.

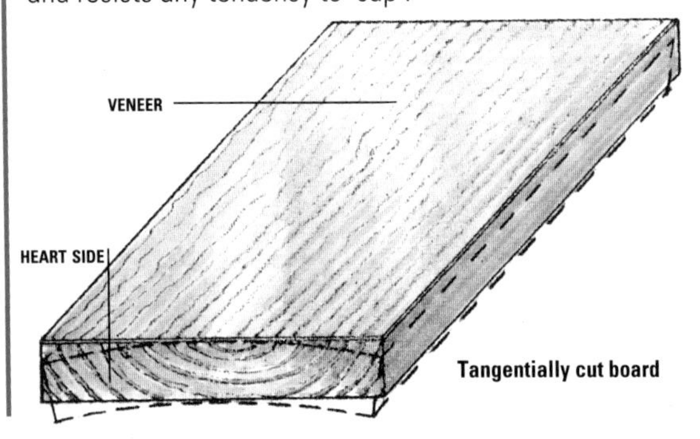

Tangentially cut board

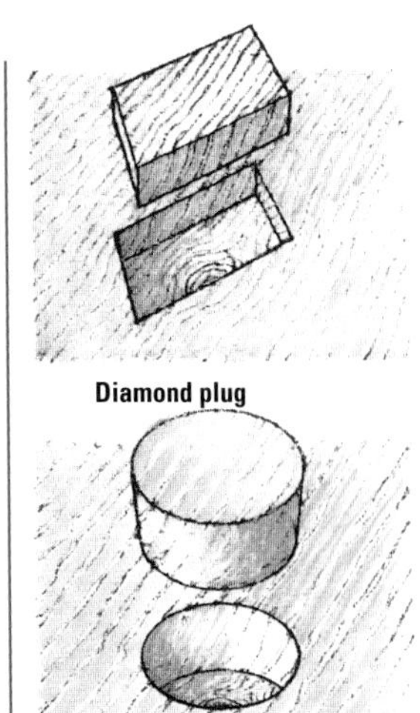
Diamond plug

Round plug

Repairing defects

Although defect-free solid wood should be selected wherever possible, unavoidable defects such as fine knots can be cut out. The holes can be filled with diamond- or round-shape plugs in which the grain follows that of the wood. If the plugs are made slightly thicker than the board, they can be levelled with a plane after gluing.

Preparing laminated boards

Other than being cut to size, most man-made boards are supplied ready for use. The surface of laminated boards must be keyed and sized; if the grain direction of the final veneer runs in the same direction as the board, an intermediary veneer that runs across the board must be applied first.

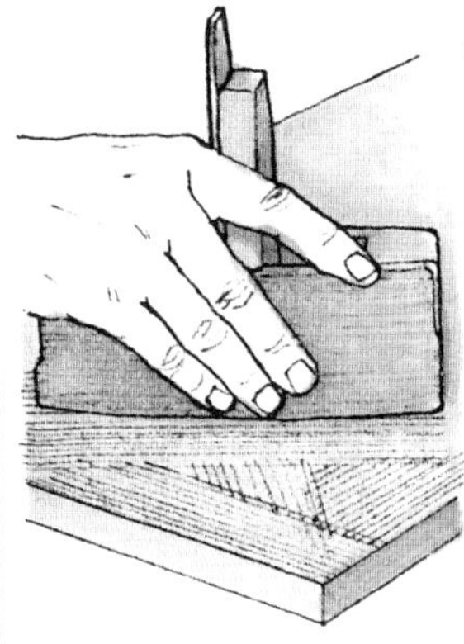

Keying the surface

For good glue adhesion, the surface of solid wood or laminated-wood groundwork must be keyed, using either a toothing plane worked diagonally from opposite sides, or a tenon saw dragged across the surface. Loose dust is vacuumed off before sizing.

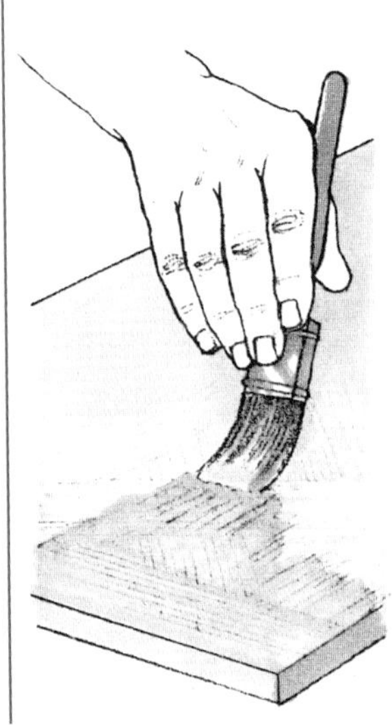

Sizing the surface

Size is made by diluting hot animal glue (see page 100) with water, at a ratio of about one part glue to ten parts water; cold wallpaper paste can also be used. The size is applied evenly to the keyed surface, and the edges are liberally coated. The rate at which size is absorbed varies according to the type of board used. When the size is thoroughly dry, the surface is sanded lightly to remove any nibs.

SHAPED GROUNDWORK

The flexibility of veneer allows it to be laid on curved surfaces; it bends easily along its grain, and can be dampened to facilitate bending across its grain on tight curves. In addition to the methods described below, shaped groundwork can be formed by laminating thick constructional veneers (see page 34).

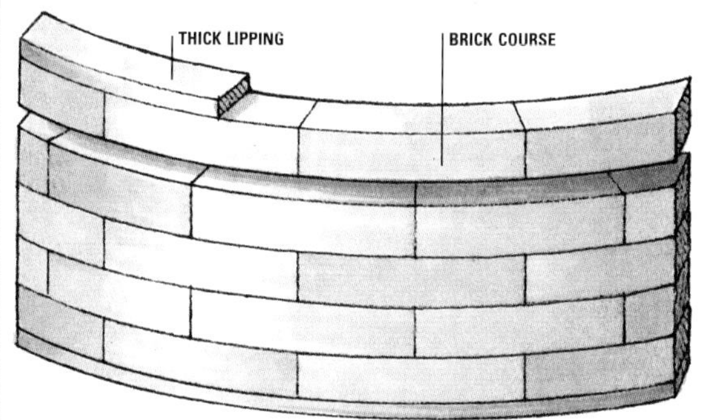

Brick construction

'Brick construction' is a traditional method of making curved groundwork, such as a bowed or serpentine drawer front, where solid-wood shapes would be wasteful; in addition, short grain can make all but the smallest components weak. In this method, the wood fibres approximately follow the direction of the curve, eliminating the problem of weak grain.

Short 'bricks' are cut from solid wood and glued end-to-end to make curved layers or 'courses'. These are staggered or 'bonded', as in conventional brickwork, so that the joints in each course are reinforced by bricks in the next. The ground is then planed and smoothed.

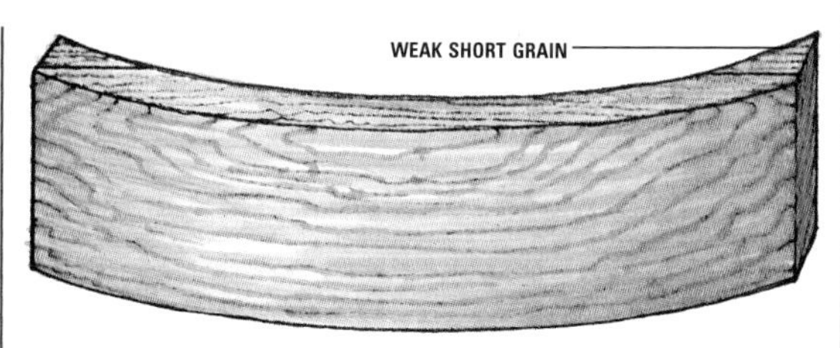

Shaping solid wood

Solid blocks of wood can be cut on a bandsaw to produce small shallow-curved groundwork. A compass plane and spokeshaves are used to smooth the curved surface, and the surface is then keyed.

Offcuts of solid wood can be faced with thick felt and used as cauls for laying the veneers (see page 103).

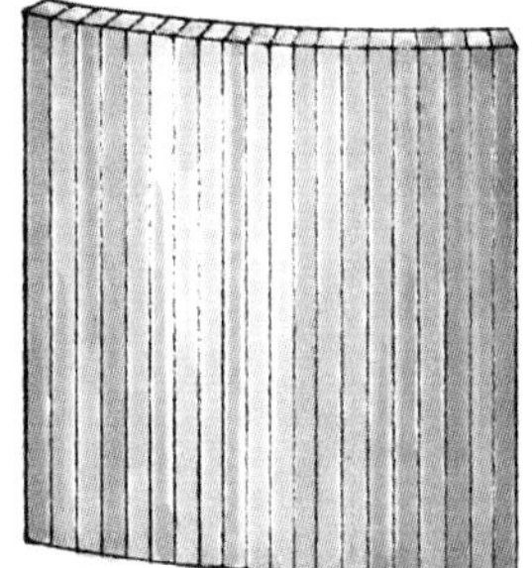

Coopered construction

Curved groundwork for larger pieces, such as bowed doors, can be constructed by gluing bevelled strips of wood edge-to-edge. The edges of each strip are planed to the required angle, and are then glued and clamped together in a specially made jig with shaped saddles to hold the curve (see bottom). When set, the surfaces are smoothed with a compass plane and keyed, ready for cross-banding veneer laid with cauls (see page 102).

The bevelled strips for smaller, lighter coopered curves can be clamped together using bands of adhesive tape.

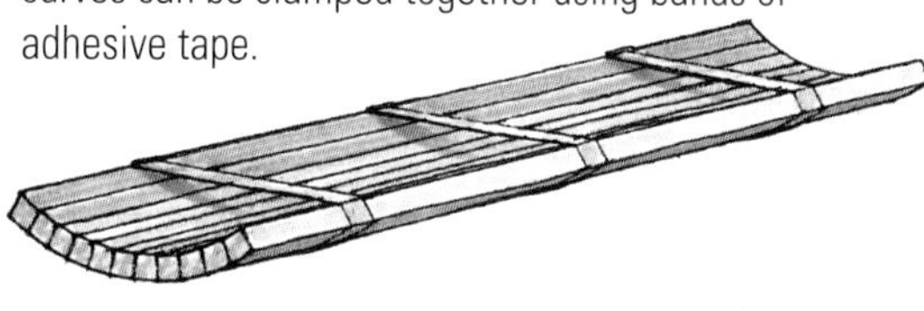

Using adhesive tape

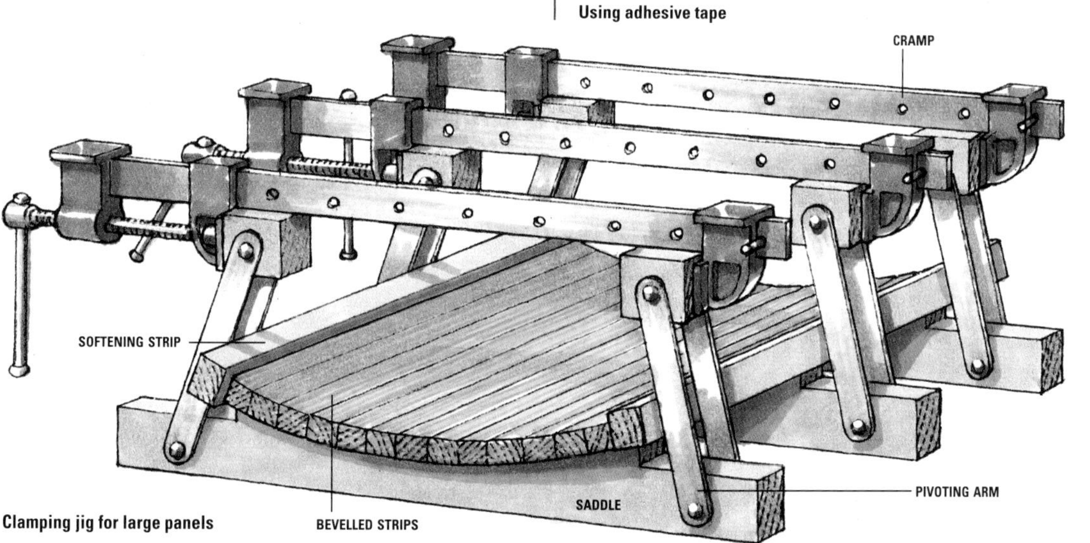

Clamping jig for large panels

PREPARING VENEER

Using veneer gives the woodworker the chance to concentrate on the decorative features of wood, as the groundwork provides all the necessary strength and structure of the workpiece.

Once the veneer has been selected, the choice mostly comes down to whether to simply utilize the natural veneer figure and colour, or to cut and manipulate the leaves to create matching patterns.

Storing and handling veneer

Because veneer is a fragile material, it must always be handled with care. This involves storing veneers flat and keeping them in the order in which they were supplied, so that they can be easily matched. Veneers should never be pulled from the middle or bottom of a stack; the leaves above those wanted should be lifted off and replaced in order. If they cannot be rolled, long veneers should be handled and moved by two people.

Flattening veneer

Most veneer will need to be flattened before it can be worked; this is best done just prior to laying.

1 Dampening the veneer
Veneer that is only slightly distorted can be moistened by steam from a kettle, immersed lightly in water or wiped with a damp sponge.

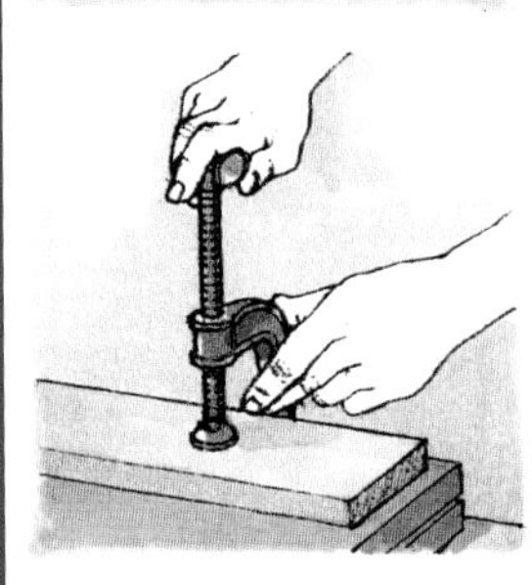

2 Pressing the veneer
The veneer is then pressed between sheets of chipboard until dry; cramps or heavy weights can be used to apply the pressure.

Using adhesive

An adhesive can be used when dampening buckled and brittle veneer. Wallpaper paste or a weak solution of hot animal glue (see page 100) is brushed lightly onto the veneer, which is then pressed between boards lined with thin polythene sheet for at least 24 hours. Heating the boards speeds up the process.

Veneer matching

Veneer that is narrower than the groundwork must be joined. The various processes of joining or matching veneer provide an opportunity to create decorative effects by juxtaposing the wood's natural features of figure and colour.

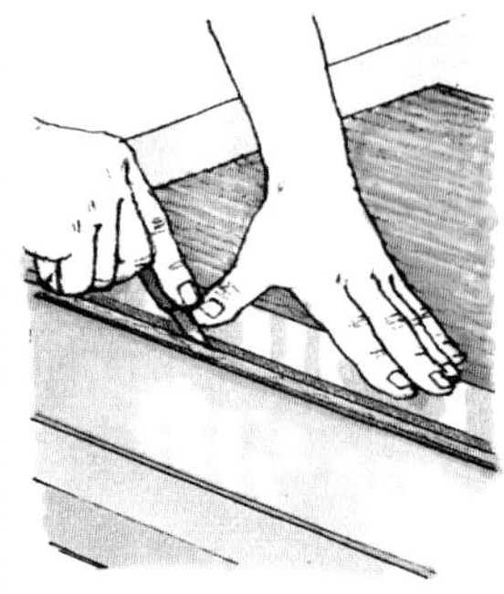

Jointing veneer
The meeting edges of the veneer must be cut straight. When two veneers are to be matched, they are laid together with the figure exactly aligned and then pinned to the cutting board temporarily. With the veneers held down with a straightedge set just inside the edge to be cut, a knife or veneer saw makes a cut through both.

The fit is checked by holding the edges together against the light. Any gaps are eliminated by 'shooting' the edges – running a finely set bench plane along the edges of the veneer set between two straight battens.

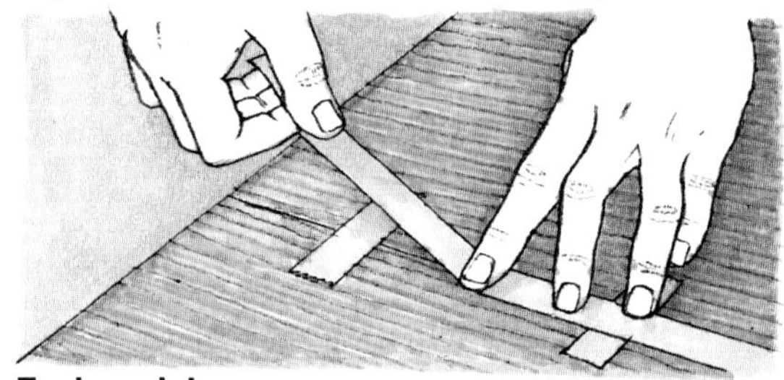

Taping a joint
The edges are placed together and lengths of veneer tape are laid about 150mm (6in) apart across the joint, before a further length of tape is laid along it. As the tape shrinks, it pulls the joint together.

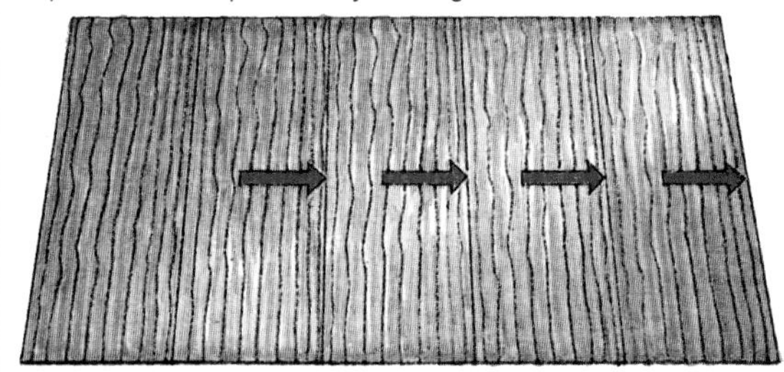

Slip matching
The simplest form of matching is used to make a wide veneer covering from narrow ones, by slipping consecutive veneers sideways and joining them together without altering their grain direction.

Slip matching is best used for striped veneers, where the joint line is not particularly obvious. Poorly matched joints, with stripes that do not run parallel to the jointing edges, must be trimmed to true up the figure.

Book matching

When the figure of two consecutive sheets of decorative veneer is biased to one side of the leaf, the veneer is usually book-matched.

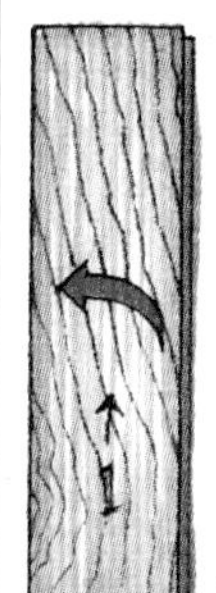
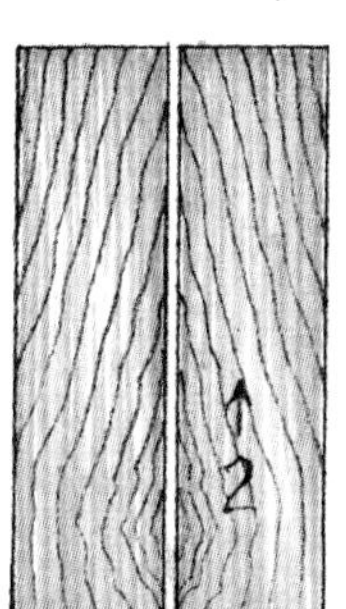

Top veneer turned left

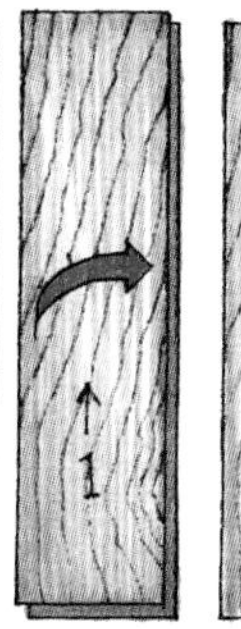
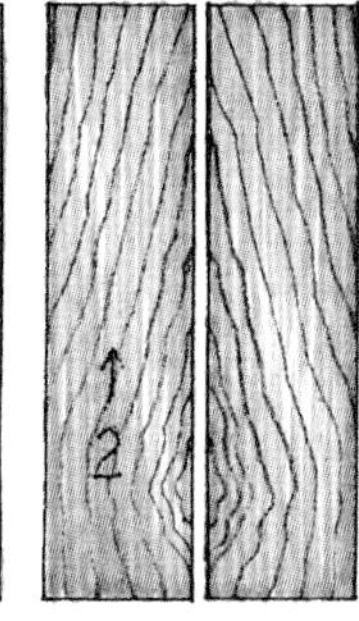

Top veneer turned right

Turning the leaves

The top leaf is turned according to the position of the dominant figure: if the figure is on the left, the leaf is turned to the left, as for opening a book, and if it is on the right, the leaf is turned to the right. Either way, the figure must be perfectly aligned to avoid a disjointed, unattractive match.

Four-piece matching

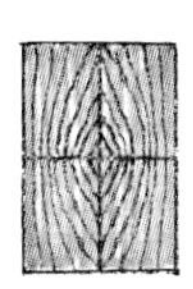

This extension of the book-matching method uses four consecutive veneers which have the focal point of the figure at the bottom.

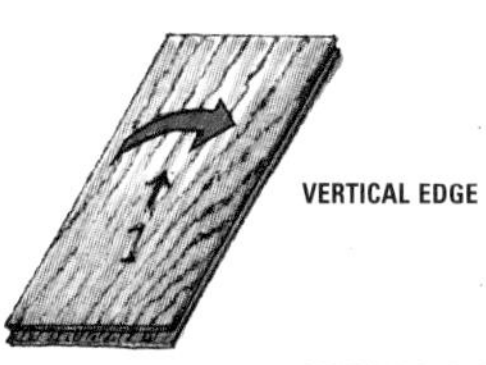

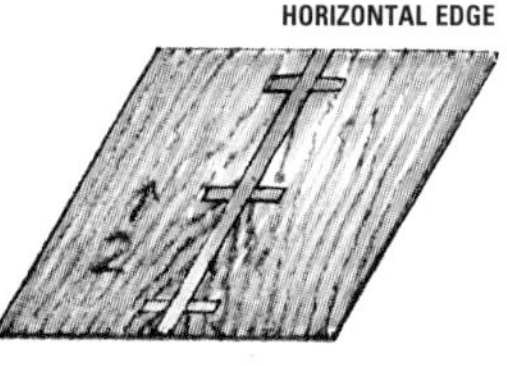

1 Taping the first pair

Each pair of leaves is book-matched in turn. The jointing edge of one leaf is trued up and laid over the edge of the adjacent leaf to match the pattern. The second leaf is cut, and the joint is matched and taped before the horizontal edge is cut square and true.

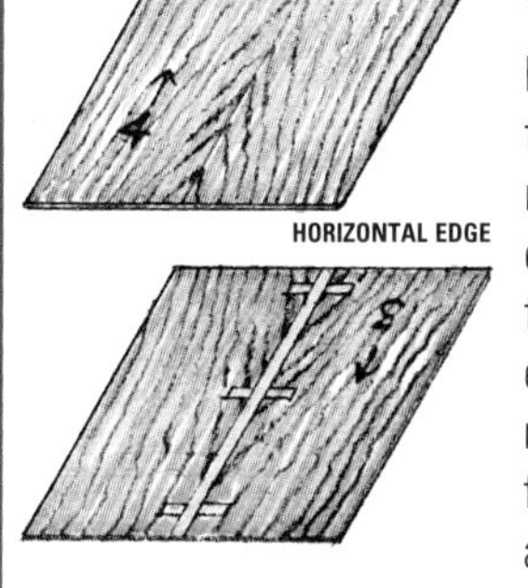

2 Taping the second pair

The second pair of leaves is book-matched and trued in the same way, and is then reversed along the horizontal edge, so that the leaves are face-down. The horizontal edges of both pairs are matched and cut when the figure pattern matches; they are then taped for laying.

Diamond matching

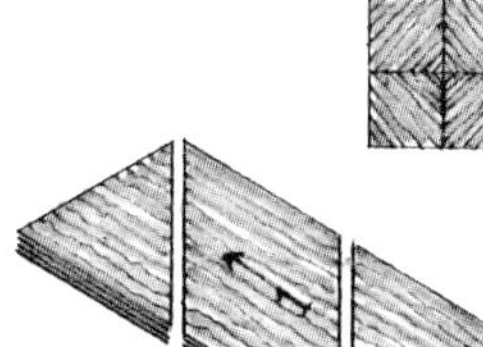

Four consecutive striped veneers are laid together and the two long edges are trued; the ends are then cut to 45 degrees, parallel to each other.

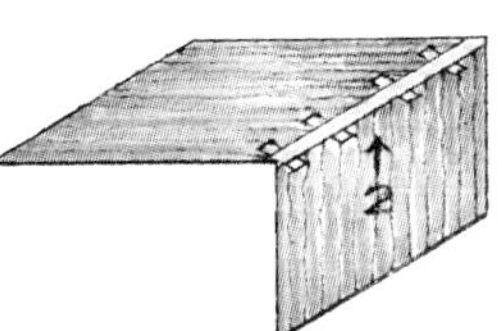

1 Making the V

The top two veneers are opened in book-match fashion and turned along the top diagonal edge to form an inverted V, which is then taped.

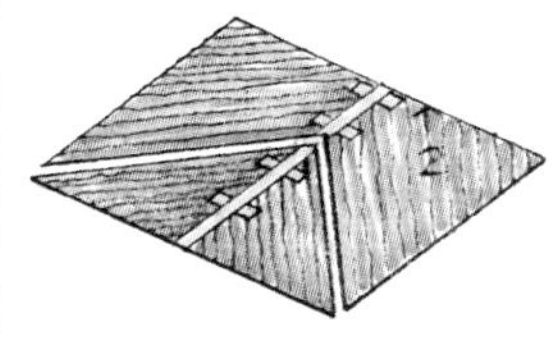

2 Making a rectangle

A straight horizontal cut is made from corner to corner. The triangular piece is fitted into the V at the bottom to form a rectangle.

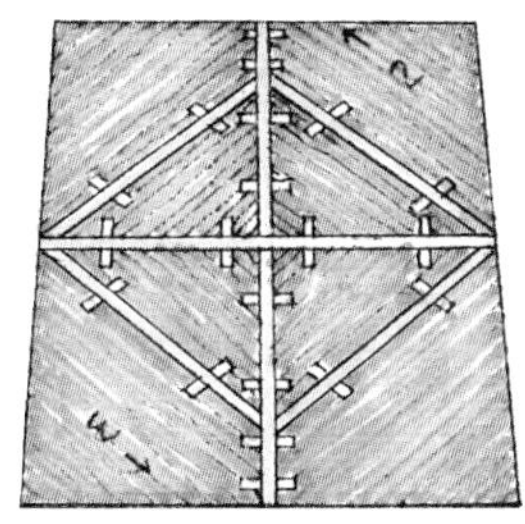

3 Joining the rectangle

The second pair of veneers is reversed so that the leaves are face-down, as for four-piece matching, and the entire process is repeated. The two rectangles are finally matched along the centre.

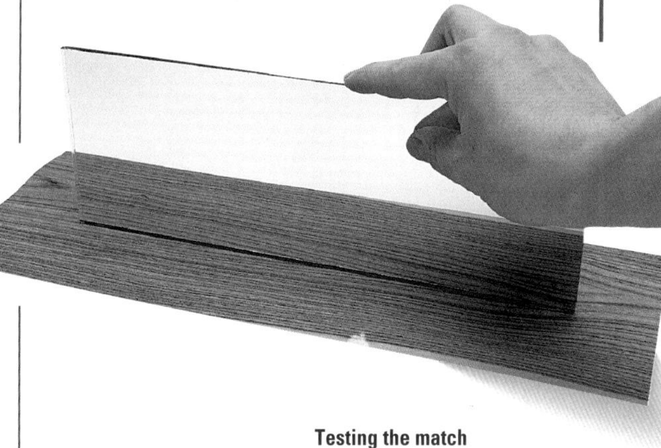

Testing the match
A mirror held perpendicular to the veneer and slid over the surface will show how a grain pattern will repeat. When the best point is reached, a cutting line is drawn along the bottom edge of the mirror; the other veneers are cut to match.

HAND VENEERING

The traditional technique of using hot animal glue as an adhesive for laying veneers by hand is still popular. Modern glues may be simpler and less messy to use, and do not require the same amount of preparation and practice; however, hot animal glue can be softened with heat, even after many years, making it a relatively simple task to correct errors and repair damaged and blistered veneer.

The sequence of work shown here is for laying veneer that covers the groundwork in one piece – matched and delicate veneers are best laid using cauls (see page 102).

Hammer veneering

Because successful hammer veneering largely depends on keeping the heated glue at a working temperature, the workplace should be kept warm and dust-free.

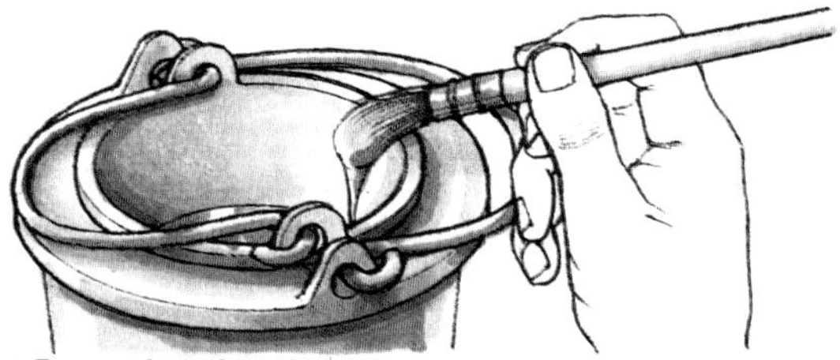

1 Preparing the glue

Animal glue, either in liquid form or as pearls which are soaked beforehand, is heated to about 49°C (120°F) in a double or jacketed glue pot (see page 95). It is stirred to a smooth, lump-free consistency that runs from the brush without separating into droplets. The glue must not be heated to boiling point, and the heated water in the outer pot or jacket must not be allowed to run dry.

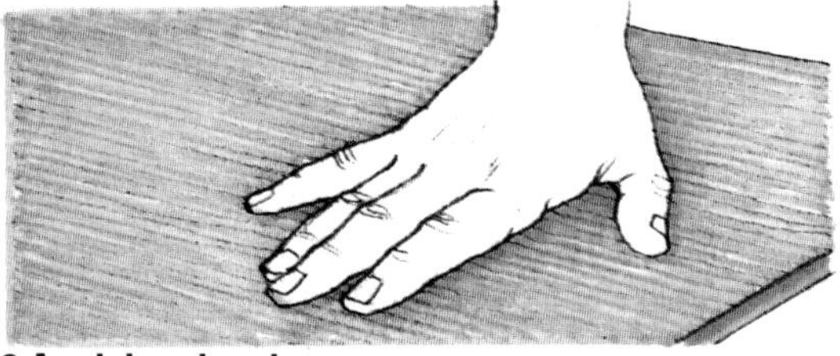

2 Applying the glue

A thin, even coat of glue is brushed onto both the groundwork and the veneer; sizing the groundwork beforehand ensures that an excess of glue is not absorbed, leaving the bond weak. The two pieces are then put aside. When the glue is almost dry but still tacky, the veneer is laid onto the groundwork, overlapping it all round, and is smoothed down with the palm of the hand.

3 Pressing the veneer

Dampening the surface of the veneer, using a sponge dipped in hot water and squeezed almost dry, closes its pores and prevents the iron sticking to it. The heated iron is run over the surface, to melt the glue and draw it into the veneer.

4 Using the hammer

The veneer hammer is used to press the veneer onto the groundwork, starting near the centre and working towards the edges with a zigzag stroke.

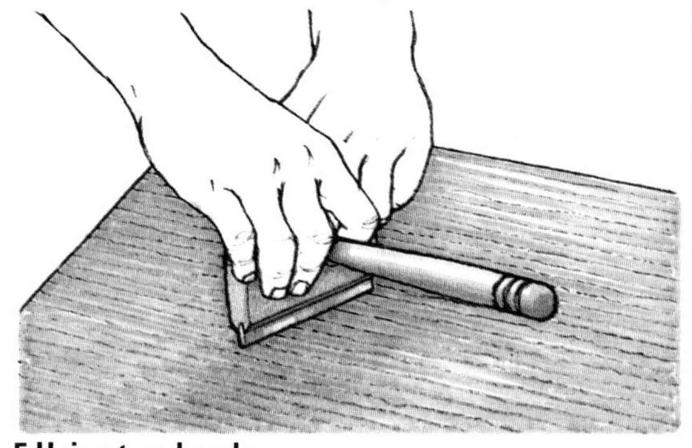

5 Using two hands

As air and melted glue are forced out from beneath the veneer, pressure on the hammer is increased by using two hands – it is important, however, not to stretch the veneer by pressing too hard across the grain.

Re-heating the glue

If the glue cools during pressing, the veneer surface should be dampened and ironed again and the pressing repeated. Melted glue must be cleaned off the surface before it sets, using a damp cloth.

CHECKING FOR BLISTERS

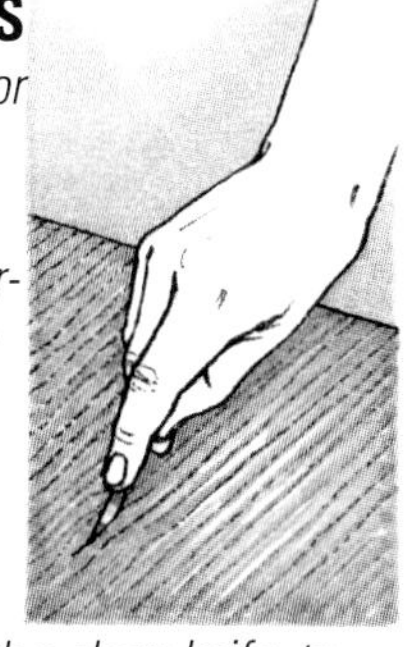

When the glue has set, blisters or air bubbles trapped beneath veneer can be detected by tapping the surface with a fingernail. Hollow-sounding areas can be treated by re-pressing the veneer, using a heated iron and veneer hammer. If this does not solve the problem, a small slit along the grain can be made with a sharp knife, to allow air to escape. The veneer is then pressed down.

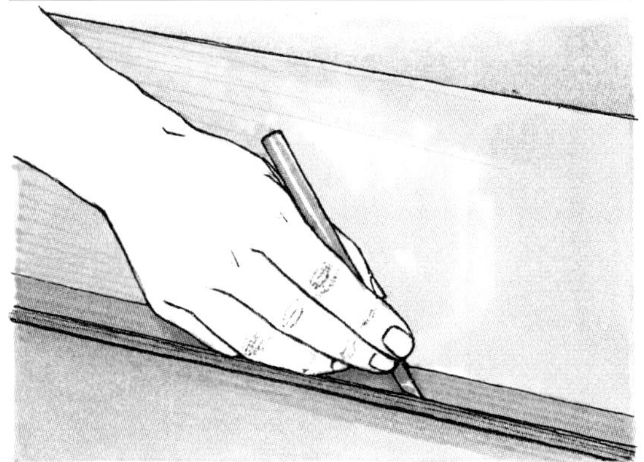

Trimming surplus veneer

When the glue has set, a trimming tool can be used to trim the edges of the panel. Alternatively, the panel can be turned face-down on a flat cutting board, so that the surplus veneer can be trimmed flush with the groundwork, using a sharp knife. To avoid splitting veneer, cross grain should be trimmed from the corners towards the centre.

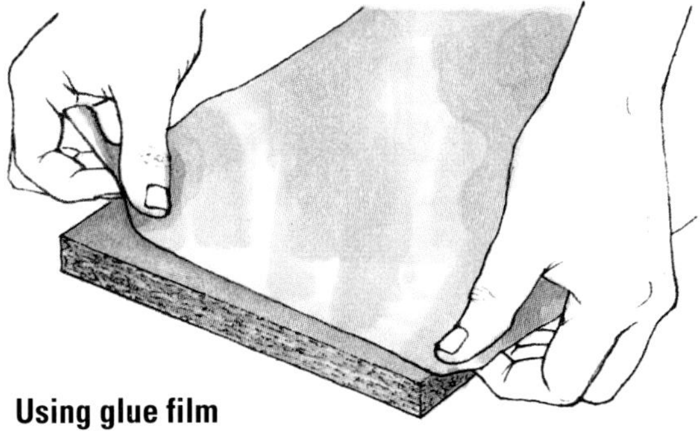

Using glue film

The modern equivalent of traditional hot animal glue is a paper-backed film of glue. It is supplied ready for use and can be reworked in the same way, with a further application of heat. Although glue film generally takes less skill to apply than animal glue, difficult veneers, such as burrs and curls, still need practice for successful application, and may be laid best by pressing with cauls (see page 102).

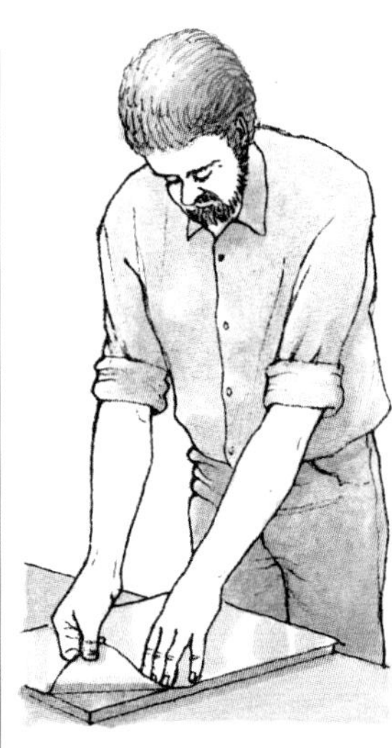

1 Applying the film

The glue film is cut with scissors to a slightly larger size than the groundwork and positioned face-down on it, before being lightly smoothed flat, using a domestic iron heated to a medium setting. The glue is allowed to cool, and the backing is then peeled off and kept.

2 Laying the veneer

The veneer is laid on the glued groundwork, and the backing paper is placed on top to protect the veneer. The heated iron is then pressed over the surface, working slowly from the centre outwards, and pressure is applied behind it, using a veneer hammer or a flat block of wood to keep the veneer pressed flat as the glue cools. Any air bubbles are removed (see top left), and surplus veneer is trimmed when the glue has set.

TONAL VARIATION

Veneer can appear lighter or darker, depending on the direction from which it is viewed. The difference in tone is apparent when consecutive veneers are laid in opposite directions.

Numbers and arrows chalked on the top face of the veneers as they are taken from the bundle help identify the grain direction and distinguish the finer 'closed' face from the slightly coarser 'open' face; ideally, the latter should be laid on the groundwork, but this is not always possible.

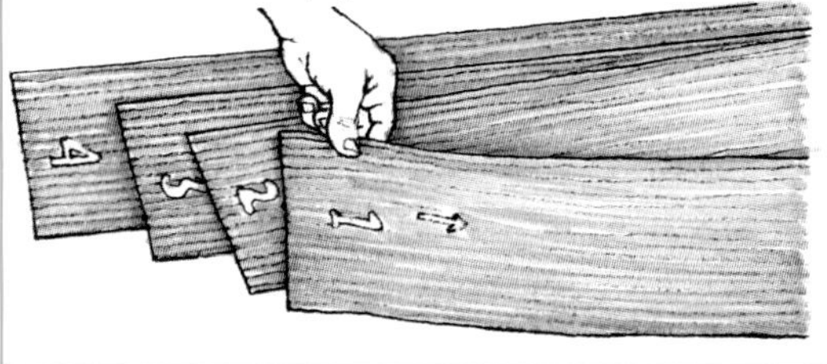

CAUL VENEERING

Cauls are flat or curved stiff boards, between which the veneer and groundwork are pressed together by clamping. Because the process means extra work and materials for making the cauls and the press, caul veneering is a more complicated job than hand veneering. It is, however, the most effective method for laying taped-together or matched veneers, or fragile veneers, such as curls and burrs, and makes it possible to veneer both sides of the groundwork simultaneously. In addition, larger curved surfaces are easier to work, as cold-setting glues allow time for 'laying up'.

Caul assemblies

Each type of caul assembly varies, according to the size and shape of the work, and the extent to which it will be used. In all caul veneering, however, the cauls must be larger than the panel to be veneered.

Small cauls

The cauls used for pressing small or narrow flat work are usually made from stout lengths of wood. The pressure is applied by cramps placed along the centre line of the cauls.

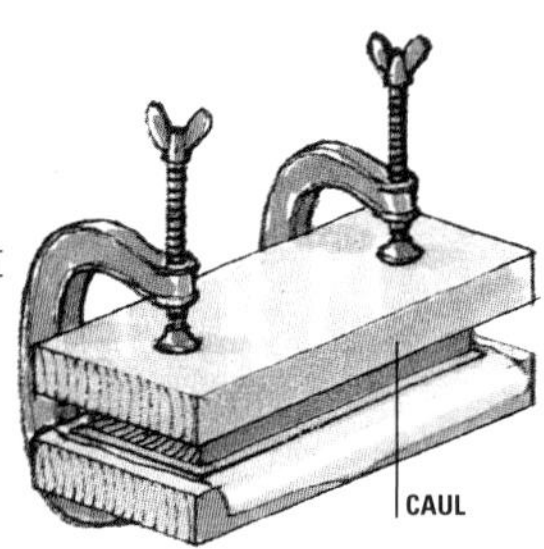

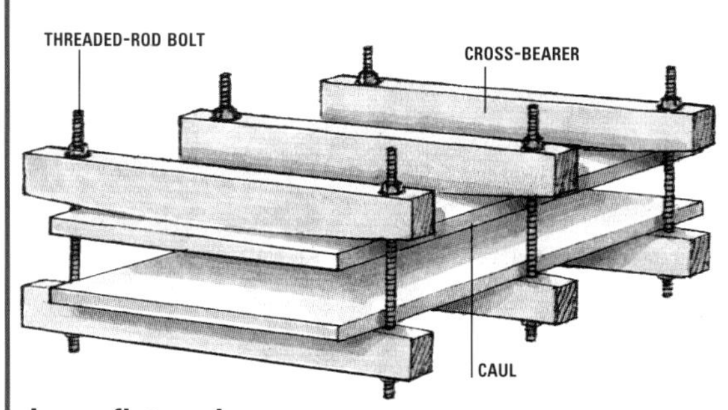

Large flat cauls

Man-made boards are used for the cauls. Pairs of stiff softwood cross-bearers, with each inner edge slightly curved, put initial pressure on the centre of the cauls, forcing surplus air and glue out. The curve compensates for the clamping forces at the ends of the bearers. The pressure can be applied by cramps, or the cross-bearers can be bolted together with threaded rods, nuts and washers. The centre cross-bearer is tightened first and then the others, working out to the ends.

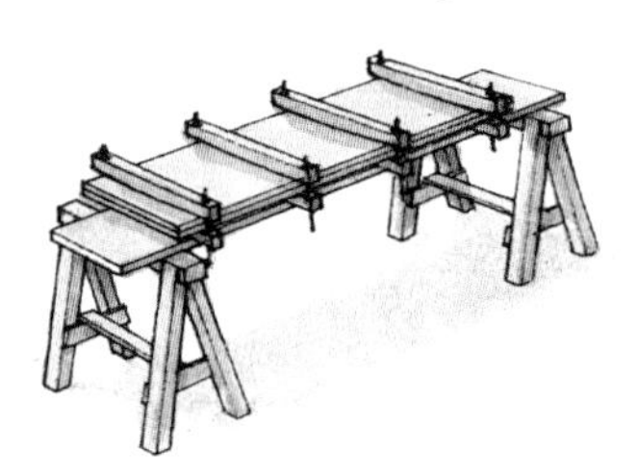

Supporting long cauls
When veneering extra-long workpieces, the caul assembly is supported by a board placed across a pair of trestles.

Preparing for work

The process of caul veneering involves a number of operations and requires everything to be ready and close at hand to save time. Where a bench is not suitable, the assembly can be set up on trestles.

Laying the veneer

If the veneer is to be laid one side at a time, the backing veneer should be pressed on first. The sequence shown here is for caul veneering both sheets at the same time.

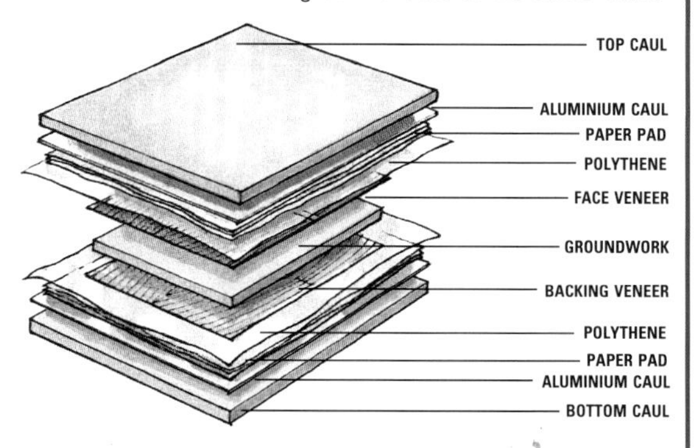

1 Working in sequence

A resin or animal-glue adhesive is brushed evenly onto the groundwork and allowed to become tacky. Glue is not usually applied to the veneer in caul veneering, to avoid the tendency for the veneer to curl. Working up from the bottom, an aluminium caul, pre-heated against a radiator or in front of a heat source, is laid on the bottom caul; this accelerates the setting time of cold-setting resin glues, and prevents premature gelling when using hot animal glues.

A pad of newspaper sheets and then a sheet of polythene are laid onto the aluminium caul; the polythene prevents the work sticking to the newspaper, which is used to take up any unevenness caused by veneer tape when pressing. The backing veneer is positioned on the polythene, and the glued groundwork is placed on it. The sequence is then built up in reverse.

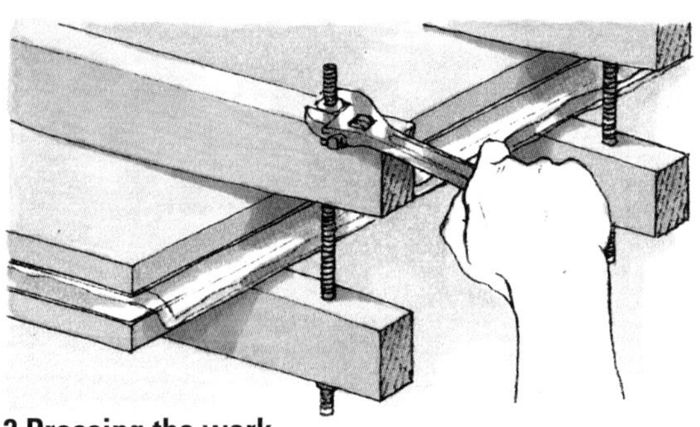

2 Pressing the work
With the top and bottom cauls in place, the whole assembly is positioned on the bottom cross-bearers. The top cross-bearers are then fitted and pressure is applied for up to 12 hours.

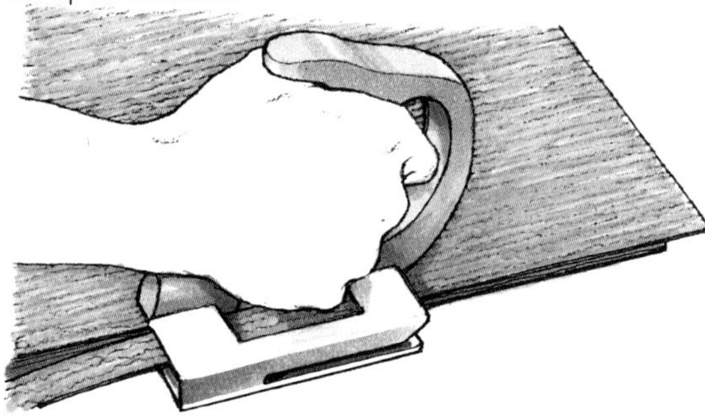

3 Finishing off
When the work is taken out of the cauls, surplus veneer is trimmed away and the board is stood on edge for a few days, to allow even air circulation, before the edges are planed and lipped (see page 116).

USING A SANDBAG

Veneer for small curved work can be pressed into place using a hot sandbag. A flat bag is filled with sand and then heated, against a radiator or in front of a heater, before being pressed into or around the shaped work and clamped with cauls and cramps.

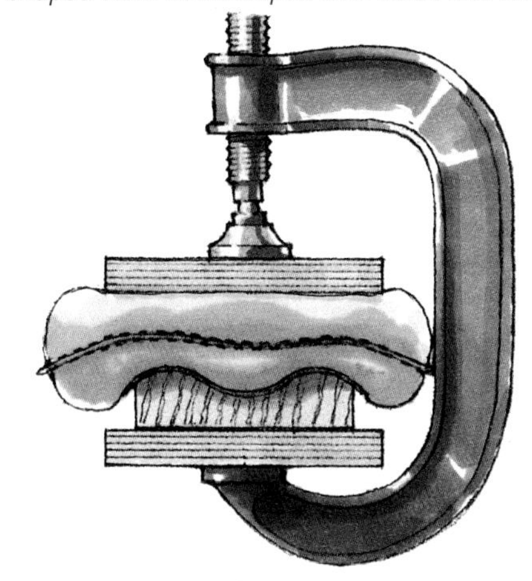

Curved cauls

Curved panels can be caul-veneered using two-part male and female formers, as for bending laminated wood (see page 35). Alternatively, the work can be glued in a press similar to those used for flat veneers, but with cauls made from strips of wood and held in shaped cross-bearers. The cauls can be made rigid, but flexible cauls offer a wider range of uses.

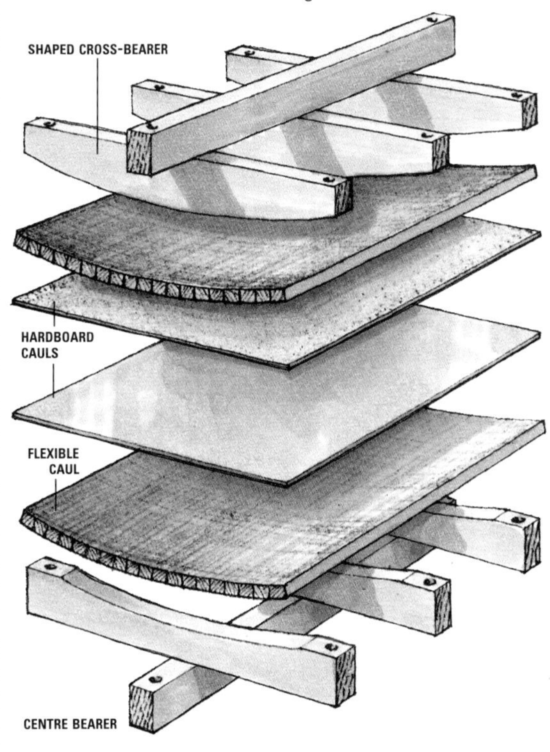

Curved-caul assemblies
The curves cut in the cross-bearers should allow for the thickness of the caul material and the groundwork. A section drawing is used to calculate the curves, and enough pairs of top and bottom cross-bearers are cut for each pair to be spaced quite closely.

Flexible cauls are made from narrow strips of wood glued to canvas sheets; they are laid between the cross-bearers, with the canvas side facing upwards, and are lined with aluminium or hardboard cauls to even out the curved surface.

Stiff bearers are placed along the centre line of the curved cross-bearers. The whole assembly is clamped and tightened in the same way as for working with flat-caul assemblies, starting with the centre bearers.

LAYING LINES AND BANDINGS

Decorative bandings and lines can be added to a veneered surface or solid wood; for the latter, it is necessary to cut a groove into a panel or a rabbet on an edge. Corners of grooves are worked and trimmed square with a chisel.

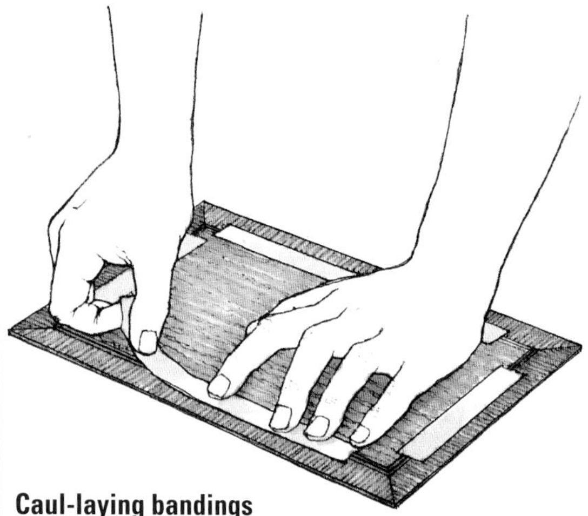

Caul-laying bandings
The centre-panel veneer is cut to size, and the decorative bandings are cut and mitred to fit around the panel. The border veneer or cross-bandings are also cut and mitred before all the parts are taped together. Glue is applied to the groundwork and the veneer is placed and set in the press.

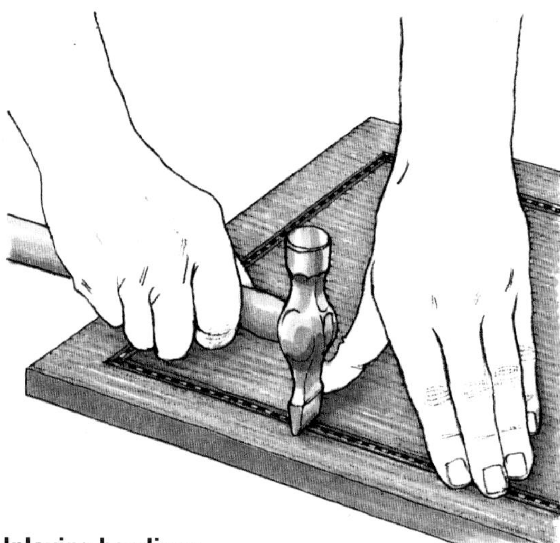

Inlaying bandings
A router is used to machine a groove around the face of the solid-wood panel, with the depth of the groove slightly less than the thickness of the banding; the corners are trimmed square with a chisel. When the inlay has been cut and mitred to fit it is glued into place and pressed down with a cross-peen hammer.

Laying cross-bandings
The centre-panel veneer is cut short of the edges of the groundwork, and laid with animal glue. When set, the veneer is trimmed parallel to the edges of the board, using a cutting gauge.

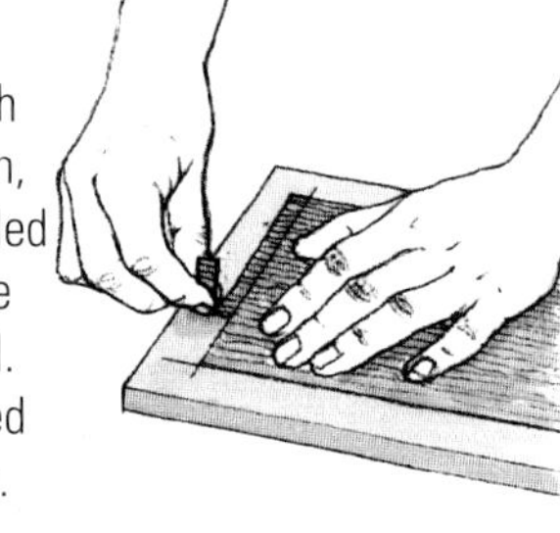

Peeling the waste
The glue is softened with heat from an electric iron, the waste veneer is peeled off and lumps of glue are scraped off with a chisel. The surface is then wiped with a warm damp cloth.

Cutting cross-bandings
Cross-bandings, cut from the ends of consecutive veneers with a cutting gauge, are then cut slightly longer and wider than required, using a straight-edged board to guide the gauge.

Gluing the bandings
Animal glue is first applied to the groundwork and both faces of the bandings. A veneer hammer or cross-peen hammer is used to lay the strips, with the ends over-lapping. The corners are mitred by cutting through both layers using a straightedge aligned with the inner and outer edges to guide the knife. The waste is removed, and the mitred ends are pressed with a hammer.

MAN-MADE BOARDS

CHAPTER 4

Man-made boards have become popular among all levels of users of wood, from the construction industry to the home woodworker, mostly for building work and furniture-making. A wide range of manufactured boards is available, and can be roughly classified into three basic types, each with its own subdivisions: laminated boards (plywood), particle boards and fibreboards.

PLYWOOD

Plywood is made from thin laminated sheets of wood called construction veneers, plies, or laminates. These are bonded at 90 degrees to each other, to form a strong, stable board; odd numbers of layers are used, to ensure that the grain runs the same way on the top and bottom.

Manufacturing plywood

A wide range of timber species of both hardwoods and softwoods is used to produce plywood. The veneers may be cut by slicing or rotary cutting (see page 90) – for softwoods, the latter is the most common method.

A debarked log is converted into a continuous sheet of veneer of a thickness between 1.5 and 6mm (1/16 and 1/4in). The sheet is clipped to size, then sorted and dried under controlled conditions before being graded into face, or core, plies. Defective plies are plugged and narrow core plies are stitched or spot-glued together before laminating.

The prepared sheets are laid in a glued sandwich, the number depending on the type and thickness of plywood required, and hot-pressed. The boards are then trimmed to size and are usually sanded to fine tolerances on both sides.

Stocks of silver birch for plywood manufacture

STANDARD SIZES

Plywood is available in a wide range of sizes. The thickness of most commercially available plywood ranges from 3mm (1/8in) up to 30mm (1 3/16in), in increments of approximately 3mm (1/8in). Thinner 'air-craft' plywood is available from specialist suppliers.

A typical board is 1.22m (4ft) wide, and boards 1.52m (5ft) wide are also available. The most common length is 2.44m (8ft), although boards up to 3.66m (12ft) can be purchased.

The grain of the face ply usually, but not always, follows the longest dimension of the board. It runs parallel to the first dimension quoted by the manufacturer, so a 1.22 x 2.44m (4 x 8ft) board will have the grain running across the width.

Plywood construction

Solid wood is a relatively unstable material, and a board will shrink or swell more across the fibres than it will along them. There is also a high risk of distortion, depending on how the board has been cut from the tree. The tensile strength of wood is greatest following the direction of the fibres, but wood will also readily split with the grain.

In order to counter this natural movement of wood, plywood is constructed with the fibres or grain of alternate plies set at right angles to one another, thus producing a stable, warp-resisting board with no natural direction of cleavage. The greatest strength of a panel is usually parallel to the face grain.

Plies

Most plywood is made with an odd number of plies to give a balanced construction, from three upwards; the number varies according to the thickness of the plies and the finished board. However many plies are used, the construction must be symmetrical about the centre ply or the centre line of the panel thickness.

The surface veneers of a typical plywood board are known as face plies. Where the quality of one ply is better than the other, the better ply is called the face and the other the back. The quality of the face plies is usually specified by a grading letter code (see opposite).

The plies that are immediately beneath, and laid perpendicular to, the face plies are known as cross-plies. The centre ply (or plies) is known as the core.

USES OF PLYWOOD

The performance of plywood is determined not only by the quality of the plies, but also by the type of adhesive used in its manufacture. Major manufacturers test their products rigorously by taking batch samples through a series of tests that exceed service requirements. The glue bond of exterior grades is stronger than the wood itself, and panels made with formaldehyde glues must comply with a formaldehyde-emissions standard. Plywoods can be broadly grouped by usage.

Interior plywood (INT)

These plywoods are used for interior non-structural applications. They are generally produced with an appearance-grade face ply and poorer quality ply for the back. They are manufactured with light-coloured urea-formaldehyde adhesive. Most are suitable for use in dry conditions, such as for furniture or wall panelling. The modified adhesive used in some boards affords them some moisture resistance, thus enabling them to be used in areas of high humidity. INT-grade plywood must not be used for exterior applications.

Exterior plywood (EXT)

Depending on the quality of the adhesive, EXT-grade plywoods can be used for fully or semi-exposed conditions, where structural performance is not required. They are often used for kitchen fitments or applications around showers and in bathrooms.

Boards suitable for fully exposed conditions are bonded with dark-coloured phenol-formaldehyde (phenolic) adhesive. This type produces weather-and-boil-proof (WBP) plywood. WBP adhesives are those which comply with an established and tested standard and have proved to be highly resistant to weather, micro-organisms, cold and boiling water, steam and dry heat over many years. Exterior-grade plywoods are also produced using melamine urea-formaldehyde adhesive. These boards are semi-durable under exposed conditions.

Marine plywood

Marine plywood is a high-quality, face-graded structural plywood, constructed from selected plies within a limited range of mahogany-type woods. It has no 'voids' or gaps, and is bonded with a durable phenolic-resin adhesive. It is primarily produced for marine use, and can be used for interior fitments where water or steam may be present.

APPEARANCE GRADING

Plywood producers use a coding system to grade the appearance quality of the face plies used for boards. The letters do not refer to structural performance. Typical systems for softwood boards use the letters A, B, C, C plugged and D.

The A grade is the best quality, being smooth-cut and virtually defect-free; D grade is the poorest, and has the maximum amount of permitted defects, such as knots, holes, splits and discolouration. A-A grade plywood has two good faces, while a board classified as B-C has poorer grade outer plies, with the better B grade used for the face and the C grade for the back.

Decorative plywoods (see page 92) are faced with selected matched veneers and are referred to by the wood species of the face veneer.

1 Trademark
The grading authority. Here, the American Plywood Association.

2 Panel grade
Identifies the grades of the face and back veneers.

3 Mill number
Code number of the producing mill.

4 Species group number
Group 1 is the strongest species.

5 Exposure classification
Indicates bond durability.

6 Product Standard number
Indicates the board meets the U.S. Product Standard.

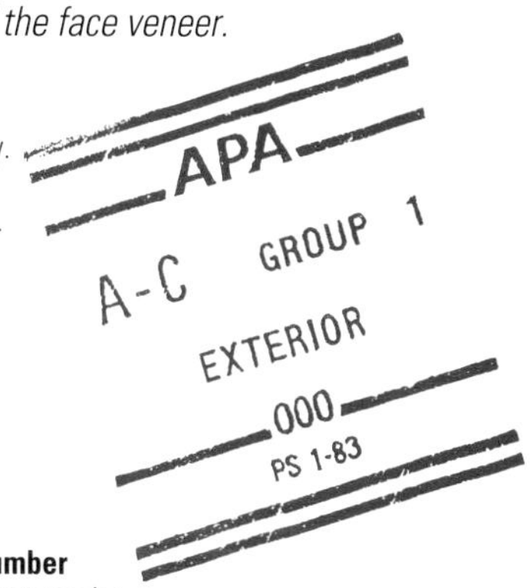

APA
A-C GROUP 1
EXTERIOR
000
PS 1-83

Stamp applied to back face

Stamp applied to edge

A-B · G-1 · EXT-APA · 000 · PS1-83

Typical grading stamps

Boards with A-grade or B-grade veneer on one side only are usually stamped on the back; those with A or B grades on both faces are usually stamped on the panel edge.

Structural plywood

Structural or engineering-grade plywood is produced for applications where strength and durability are the prime considerations. It is bonded with phenolic adhesive. A lower appearance-grade face ply is used, and boards may not have been sanded.

TYPES OF PLYWOOD

Plywood boards are manufactured in many parts of the world. The species of woods used depends on the area of origin, and the performance and suitability are affected by the species of wood, type of bond and grade of veneer.

Softwoods and hardwoods

Softwood boards are commonly made from Douglas fir or species of pine; hardwood veneers are mostly made from light-coloured temperate woods such as birch, beech and basswood. Red-coloured plywoods are made from tropical woods such as lauan, meranti and gaboon.

The face veneers and core may be made from the same species throughout, or may be constructed from different species.

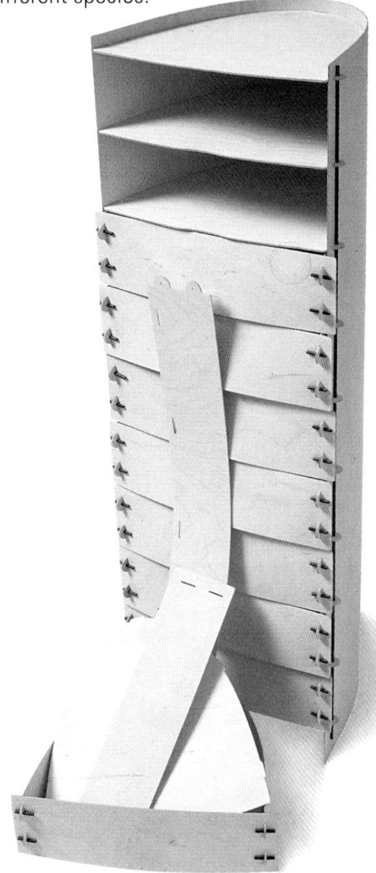

Fabricated birch-plywood chest of drawers

Applications

Different types of plywood are manufactured for such diverse applications as aircraft and marine construction, agricultural installations, building work, panelling, musical instruments, furniture and toys.

1 Decorative plywood
This is faced with selected rotary, flat-sliced or quarter-cut matched veneers, usually of hardwoods such as ash, birch, beech, cherry, mahogany or oak, sanded ready for polishing. A balancing veneer of lesser quality is applied to the back of the board. Decorative plywood is mainly used for panelling.

2 Three-ply board
The face veneers are bonded to a single core veneer. Each thickness of veneer may be the same, or the core may be thicker to improve the balance of construction. This type is sometimes called balanced, or solid-core, plywood. Composite laminated boards, which use a layer of reconstituted wood for the core, are used in the building industry.

3 Drawerside plywood
The exception to the cross-ply construction method, this has the grain of all the plies running in the same direction. It is made of hardwood to a nominal thickness of 12mm (½in), and is used in place of solid wood for making drawer sides.

4 Multi-ply
This has a core consisting of an odd number of plies. The thickness of each ply may be the same, or the cross-plies may be thicker, which helps give the board equal stiffness in its length and width. Multi-ply is much used in making veneered furniture.

5 Four-ply
Four-ply board has two thick-cut plies bonded together, with their grain in the same direction and perpendicular to the face plies. This type is stiffer in one direction and is used mainly for structural work.

6 Six-ply
Six-ply is similar to four-ply in construction, but has the core parallel to the face, with cross-ply in between.

Left to right:
Decorative, three-ply, drawerside, multi-ply, solid-core multi-ply, six-ply.

BLOCKBOARD AND LAMINBOARD

Blockboard is a form of plywood, by virtue of having a laminated construction. Where it differs from conventional plywood is in having its core constructed from strips of softwood cut approximately square in section; these are edge-butted but not glued. The core is faced with one or two layers of ply on each side.

Laminboard is similar to blockboard, but the core is constructed with narrow strips of softwood, each about 5mm (3⁄16in) thick; these are usually glued together. Like blockboard, laminboard is made in both three- and five-ply construction. Its higher adhesive content makes laminboard denser and heavier than blockboard.

Laminboard
Because the core is less likely to 'telegraph' or show through, this is superior to blockboard for veneer work. It is also more expensive. Boards of three- and five-ply construction are produced. With the latter, each pair of thin outer plies may run perpendicular to the core. Alternatively, the face ply only may run in line with the core strips.

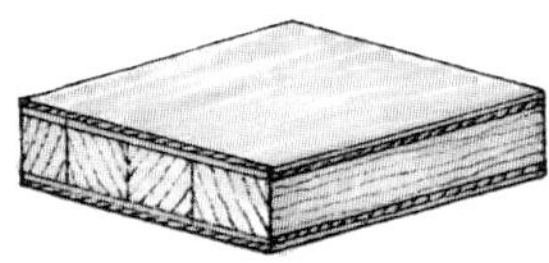

Blockboard
This stiff material is suitable for furniture applications, particularly shelving and worktops. It makes a good substrate for veneer work, although the core strips can telegraph. It is made in similar panel sizes to plywood, with thicknesses ranging from 12mm (½in) to 25mm (1in). Boards of three-ply construction are made up to 44mm (1¾in) thick.

Laminboard **Blockboard**

BENDING LAMINATED BOARDS

Thin sheets of board made with plies of the same thickness can be bent into a curve. The tightness of the bend is dependent on the board's thickness and the direction of the grain of the face veneer; it will bend more across the grain than along it. Plywood can be bent dry, but wetting the board will allow it to take up a tighter curve. To produce a very tight curve or bend a thick panel, the back of the board can be 'kerfed' – partly cut through with a row of parallel saw cuts. This effectively reduces the thickness of the board to allow it to bend; however, because it also weakens the board, it is mainly used for its visual effect.

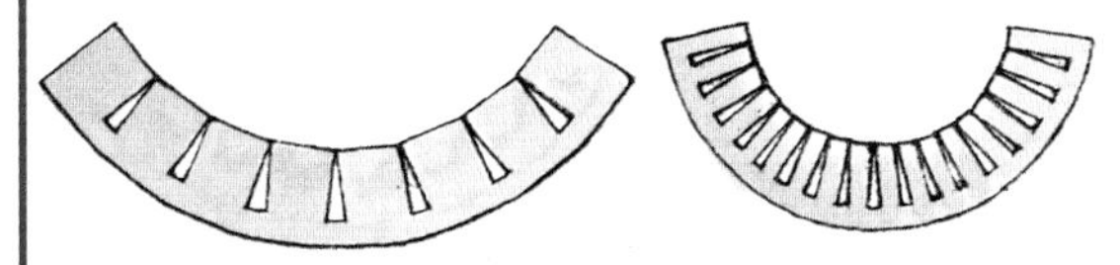

Width and spacing

The width of the kerf depends on the thickness of the saw blade: a fine saw will give a narrow cut, a coarse blade a wide cut. The width of the cut and the spaces between them determine the radius of the bend. The closer the kerfs, the tighter the bend and the smoother the curve. This form of bend, however, usually shows a faceted surface that will require sanding to final shape.

Using a power saw

Radial-arm sawing is the most accurate and quickest method of kerfing. The saw is set up for depth of cut and the distance between cuts is marked on the back fence. The first cut is made perfectly square across the panel, then aligned with the mark to give the spacing for the next cut. For larger panels, a portable power saw can be used – the cut lines are marked out and a batten is clamped to the board to guide the saw's base plate.

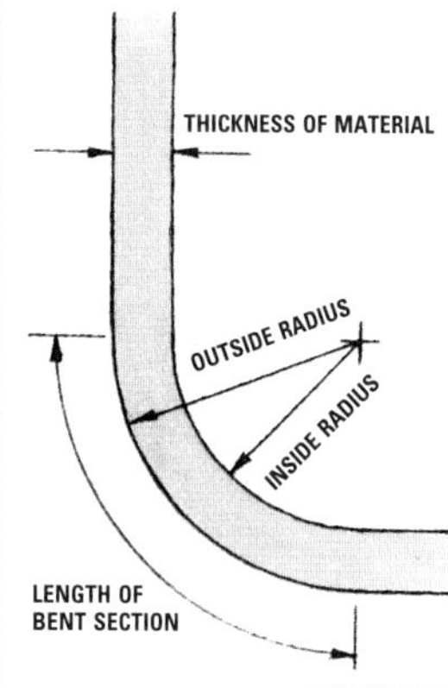

Calculating the kerf spacing

A full-size plan of the bend is drawn to determine the length of the bent section. This can be calculated or, for an approximation, measured directly from the drawing by stepping off with a pair of compasses or bending a rule around the outside curve.

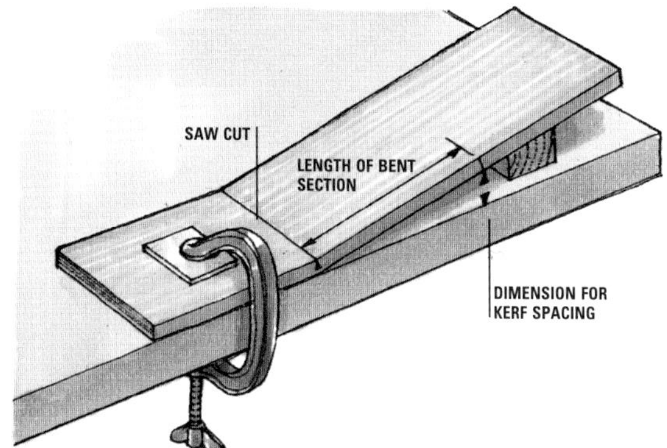

Determining the spacing

To establish the start of the bend, a saw cut is made almost through the thickness of a piece of the board, leaving at least 3mm (⅛in) of material. From this point the length of the bent section is marked off on the edge. The end of the board is clamped to a bench and the free end is lifted to close the saw cut. The distance between the underside of the board and the bench top beneath the mark gives the dimension for the kerf spacing. The actual board is then marked and kerfed to this setting.

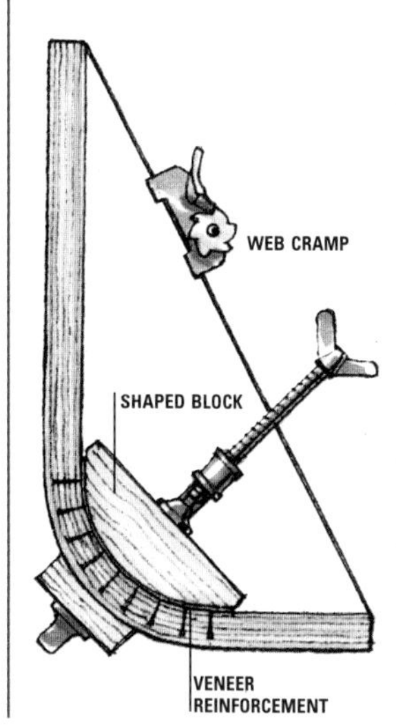

Gluing the bend

The curve is tested against the full-size drawing, using a web cramp to help hold the shape. With the board laid flat, adhesive is applied to the kerfs and the inside face. To reinforce the bend, a prepared length of veneer is bonded on the inside of the curve, with the grain perpendicular to the kerfing. The veneer is clamped with blocks shaped to fit the curve, and the bend is held with web cramps.

PARTICLE BOARDS

Wood-particle boards are made from small chips or flakes of wood bonded together under pressure – softwoods are generally used, although a proportion of hardwoods may be included. Various types of boards are produced according to the shape and size of the particles, their distribution through the thickness of the board and the type of adhesive used to bind them together.

Manufacturing particle boards

The production of particle boards is a highly controlled automated process. The wood is converted into particles of the required size by chipping machines. After drying, the particles are sprayed with resin binders and spread to the required thickness with their grain following the same direction. This 'mat' is hot-pressed under high pressure to the required thickness, and then cured. The cooled boards are trimmed to size and sanded.

Working features

Particle boards are stable and uniformly consistent. Those constructed with fine particles have featureless surfaces and are highly suited as groundwork for home veneering. Decorative boards, pre-veneered with wood, paper foil or plastic laminates, are also available. Most particle boards are relatively brittle and have a much lower tensile strength than plywood boards.

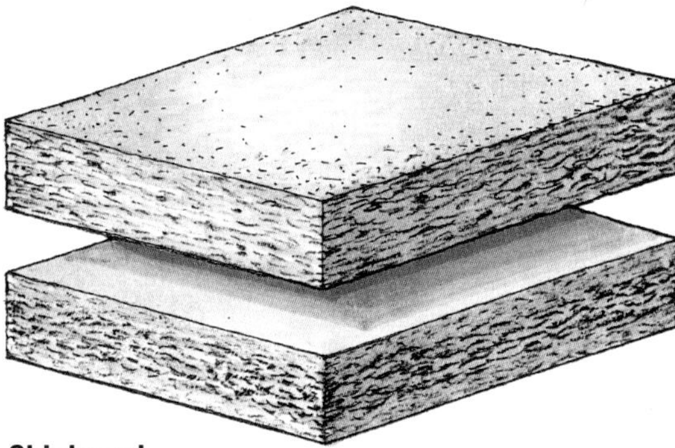

Chipboard

The types of particle board most used by woodworkers are those of interior quality, which are commonly known as chipboard. Like other wood products, interior-grade chipboard is adversely affected by excess moisture – the board swells in its thickness and does not return to shape on drying. Moisture-resistant types, suitable for flooring or wet conditions, are available, and are used extensively in the building-construction trade.

Single-layer chipboard
Made from a mat of similar-size, evenly distributed particles, this chipboard has a relatively coarse surface suitable for wood veneer or plastic laminate, but not for painting.

Three-layer chipboard
This board has a core layer of coarse particles sandwiched between two outer layers of fine high-density particles. The high proportion of resin in the outer layers produces a smooth surface that is suitable for most finishes.

Graded-density chipboard
This chipboard is a blend of coarse and very fine particles. Unlike three-layer chipboard, there is a gradual transition from the coarse interior through to the fine surface.

Decorative chipboard
These boards are faced with selected wood veneers, plastic laminates or thin melamine foil; veneered boards are sanded for polishing, and laminated boards are supplied ready-finished. Some plastic-laminate boards for worktops are made with finished profiled edges, while matching edging strips are available for melamine-face and wood-veneer boards.

Oriented-strand board
This is a three-layer material made with long strands of softwood. The strands in each layer are laid in one direction, and each layer is perpendicular to the next in the same manner as plywood.

Flakeboard or waferboard
This type of board incorporates large shavings of wood which are laid horizontally and overlap one another. Flakeboard has greater tensile strength than standard chipboard; although made for utilitarian applications, it can be used as a wallboard when finished with a clear varnish. It can also be stained.

FIBREBOARDS

Fibreboards are made from wood which has been broken down to its basic fibre elements and reconstituted to make a stable, homogeneous material. Boards of various density are produced according to the pressure applied and the adhesive used in the manufacturing process.

Top to bottom:
Oak-veneered MDF, medium-density fibre-board (MDF), low-density (LM) board, high-density (HM) board.

Top to bottom:
Perforated hardboard, decorative-face hardboard, embossed hardboard, tempered hardboard, standard hardboard.

STORING BOARDS

To save space, man-made boards should be stored on their edges. A rack will keep the edges clear of the floor and support the boards evenly at a slight angle; to prevent a thin board bending, support it from beneath with a thicker board.

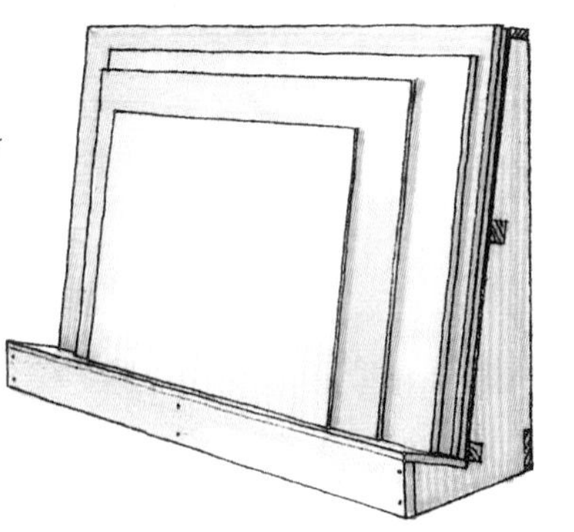

Grades of fibreboard

Most boards are made by forming wet fibres into a mat and bonding them, generally using the wood's natural resins, to produce boards of varying density.

Medium-density fibreboard (MDF)
This is manufactured using a dry process, in which synthetic resin bonds the fibres together, to produce greater strength. MDF is smooth on both faces and uniform in structure; it has a fine texture. It can be worked like wood and used as a substitute for solid wood in some applications, such as furniture-making. It can be cleanly profile-machined on the edges and faces, but does not accept screws well on its edges, which are likely to split; although stable, it swells when moist. Waterproof MDF is made for use in damp conditions. Boards of MDF are made in thicknesses of 6 to 32mm (¼ to 1¼in) and in a wide range of sizes. They make an excellent ground-work for veneer, and take paint finishes well.

Low-density (LM) board
This relatively soft board, usually 6 to 12mm (¼ to ½in) thick is used for pinboard or wall panelling.

High-density (HM) board
Heavier and stiffer than LM board, this is used for interior panelling applications.

Hardboards

Hardboard is a fibreboard that is manufactured in a similar way to LM and HM boards, but at higher pressures and temperatures.

Standard hardboard
This has one smooth and one textured face, and is made in a range of thicknesses, commonly from 3 to 6mm (⅛ to ¼in) in a wide range of panel sizes. It is an inexpensive material, commonly used for drawer bottoms and cabinet backs.

Duo-faced hardboard
Made from the same material as standard board, this has two smooth faces. It is used where both faces are likely to be seen, such as a panel of a framed door or cabinet.

Decorative hardboard
This is available as perforated, moulded or lacquered boards. Perforated types are used for screens, most others for wall panelling.

Tempered hardboard
Standard hardboard is impregnated with resin and oil to produce a stronger material that is water- and abrasion-resistant.

WORKING MAN-MADE BOARDS

Although man-made boards are relatively easy to cut, using woodworking hand tools and machines, the resin content in the boards can quickly dull cutting edges; tungsten-carbide-tipped (TCT) circular-saw blades and router cutters will keep their edge longer than standard steel ones.

The boards can be awkward to handle, due to their size, weight or flexibility. Cutting a board into smaller sections requires clear space with adequate support, and possibly the help of an assistant.

Cutting by machine
Clean-cutting high-speed machine tools will give the best results when cutting man-made boards, but will dull quickly in the process. A universal saw blade with tungsten-carbide tipped teeth should be utilized for cutting a large amount of board. The board should be face-down when using a hand-held power saw, and face-up for a table saw.

Cutting by hand
A 10 to 12PPI panel saw should be used for hand-sawing; a tenon saw can be employed for smaller work. In either case, the saw should be held at a relatively shallow angle. To prevent break out of the surface when severing fibres or laminate, all cutting lines should be scored with a sharp knife.

Threaded inserts

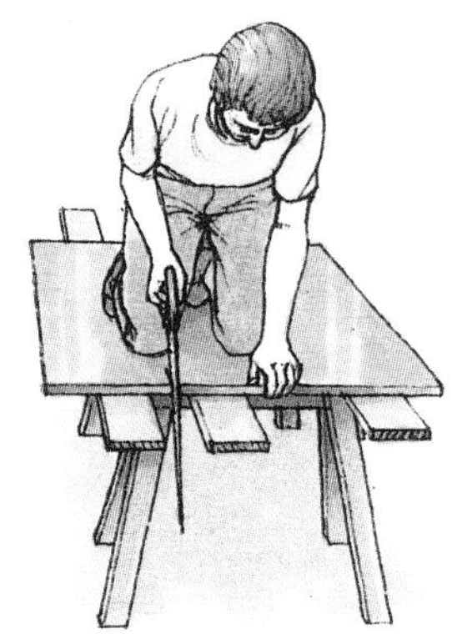

Supporting the board
The board should be supported close to the cut line and laid over the bench; stout boards can be supported on planks between trestles.
Large boards can be climbed upon, in order to reach the cutting line comfortably; an assistant can support the offcut if it is unmanageable. The solo woodworker can saw between the planks or set up some means of supporting the offcut, to prevent it breaking away before the cut is completed.

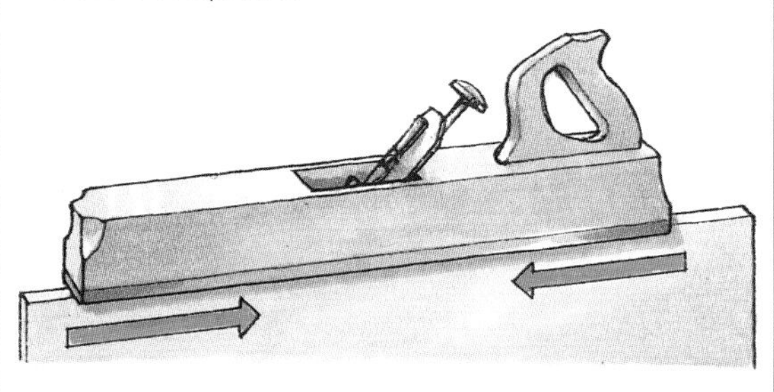

Planing the edge
Edges are planed in the same way as solid wood, but each edge is planed from both ends towards the middle, as with end grain, to prevent break out of the core or surface veneers. The blade will need to be sharpened regularly during the course of planing.

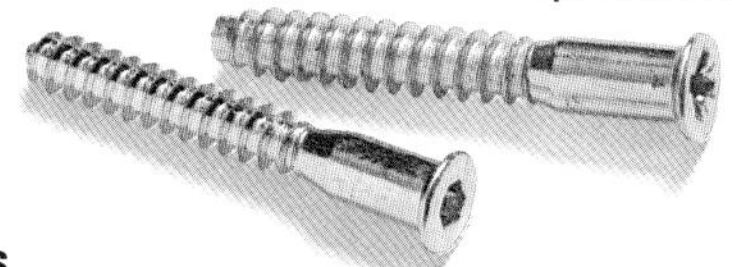

Chipboard screws

Fixing boards
Screw fixings in the edge of man-made boards are not as strong as those in the face. Pilot holes should be drilled in the edge of plywood to prevent splitting. The diameter of the screws should not exceed 25 per cent of the board's thickness.

In chipboard, screw-holding is dependent on the density of the board; most boards are relatively weak. Special chipboard screws are better than standard woodscrews. Pilot holes must always be drilled for both face or edge fixings, and special fastenings or inserts can be used for improved holding.

Blockboard and laminboard will hold screws well in the side edges, but not in the end grain.

LIPPING BOARDS

The edges of man-made boards are usually finished with a lipping to cover the core material. Long-grain or cross-grain veneer, or a more substantial solid-wood lipping of matching or contrasting wood, can be used. Lippings can be applied before or after surface-veneering.

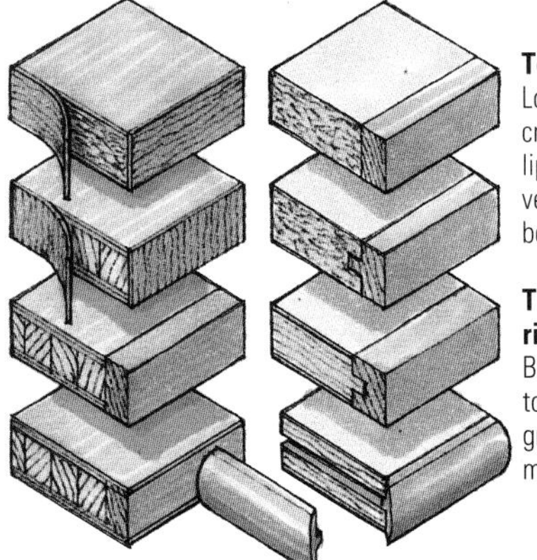

Top to bottom, left: Long-grain veneer, cross-grain veneer, lipped after veneering, lipped before veneering.

Top to bottom, right: Butt-jointed, tongued lipping, grooved lipping, mitred lipping.

Iron-on lipping

The simplest edging to apply is the pre-glued veneer type, which is ironed onto the edge of the board. These edge lippings are primarily sold for finishing veneered chipboard and are available in a limited range of matching veneers.

Solid-wood lipping

A more substantial edging which can be shaped is made from thick lippings cut from matching wood. The lipping can be butt-jointed or, for greater strength, tongued and grooved to the edge. Mitred corners improve the final appearance, particularly when the edges are moulded.

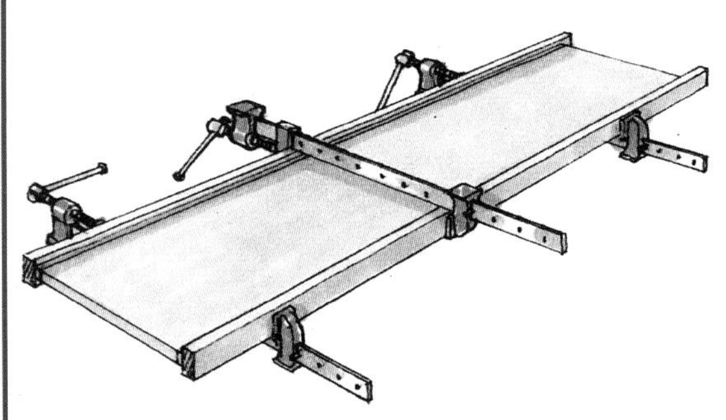

Gluing lippings

The end grain should be sized well first. When gluing a long lipping, a stiff batten should be held between the cramp heads and the work. This helps spread the clamping forces over the full length of the lipping and prevents the cramp stocks damaging the lipping.

SHAPES FOR CORNER LIPPINGS

In each of the examples shown below, the lipping forms a corner joint (see opposite) and also masks the core of the board.

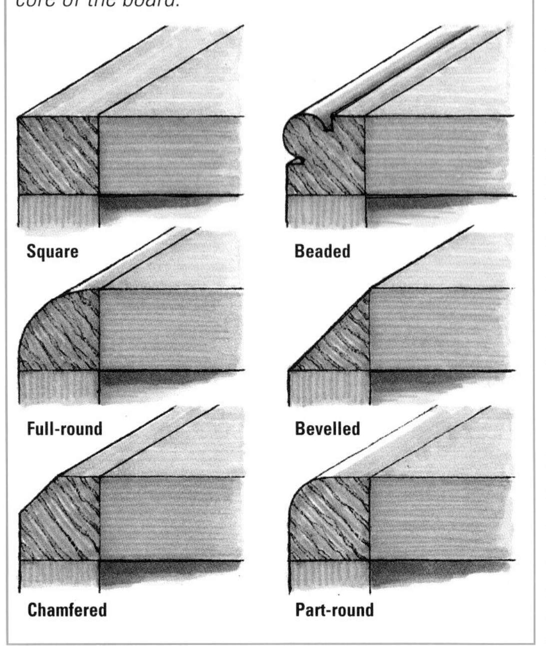

Planing lippings

When planing the width of the lipping, care must be taken to avoid damaging the surface veneer, particularly when working across the direction of the grain. The edge can be finished using a sanding block.

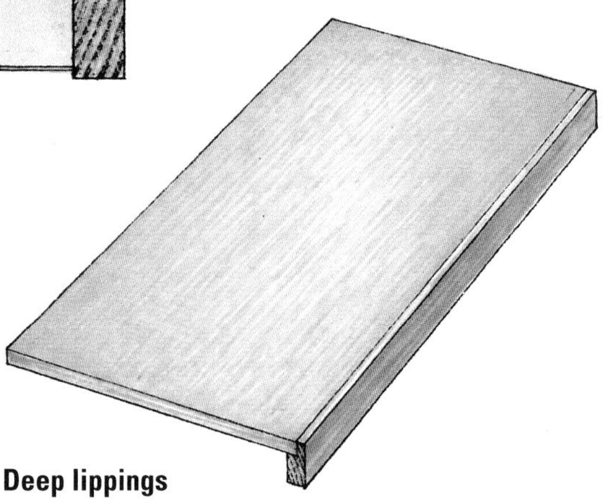

Deep lippings

Deep lippings will substantially stiffen boards for use as shelving or worktops. The full thickness of the board can be let into a rabbet cut in the lipping, or can be tongued and grooved.

BOARD JOINTS

Plywood, blockboard, laminboard, chipboard and medium-density fibreboard (MDF) can be used for carcass construction. Although man-made boards are more stable than panels of solid wood, on the whole they do not possess its long-grain strength. The means of joining these boards varies according to their composition; most joints employed in solid-wood carcass construction can be used, but framing joints, such as mortise and tenon, half laps and bridle joints, are unsuitable.

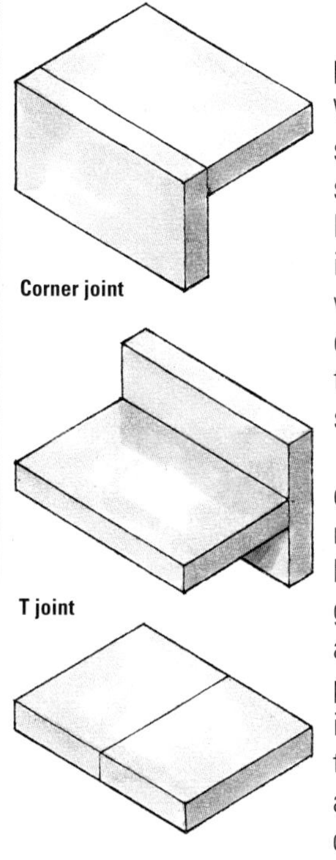

Machined joints
When selecting a joint, solid-core laminated boards such as blockboard and laminboard can be regarded in the same way as solid wood. A dovetail, for example, would be cut in the end grain, but not in the side grain.

Dovetails are quite difficult to cut by hand in man-made materials, because of the changing grain direction of the core, and machined dovetails are preferable. Most joints, including rabbet, housing, tongue-and-groove, dowel and biscuit joints, are best cut with machines.

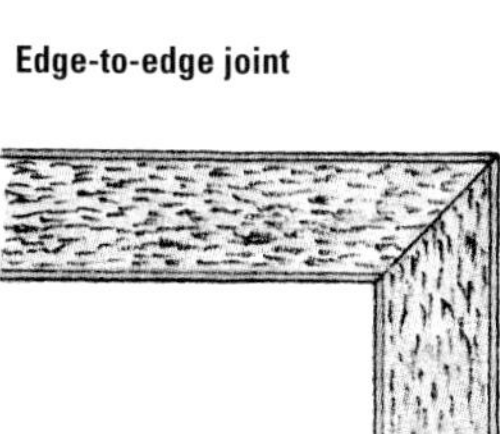

Mitred joints
Man-made boards which have been ready-finished with a decorative veneer must be mitre-jointed if the core is not to show on the face. The corner can be simply glued or reinforced using a loose spline or a fixed tongue.

Joints for corner lipping
For pre-veneered chipboard, a lipping can be used to make a corner joint which also masks the core. The lipping can be left square or shaped (see opposite), and contrasting wood can be used as a decorative feature. The lipping can be simply butt-jointed, or can be fitted with a stopped loose tongue-and-groove joint, which will provide greater structural strength.

Using a thicker lipping
A thicker lipping can be used to make a stronger joint suitable for plinths or carcass construction; in this process, a barefaced tongue is cut on the board and a matching groove in the lipping. The section can be shaped if required, as indicated by the broken line.

Lipping chipboard
For greater strength when using chipboard, a groove can be cut in the edge of the board and a tongue formed on the lipping.

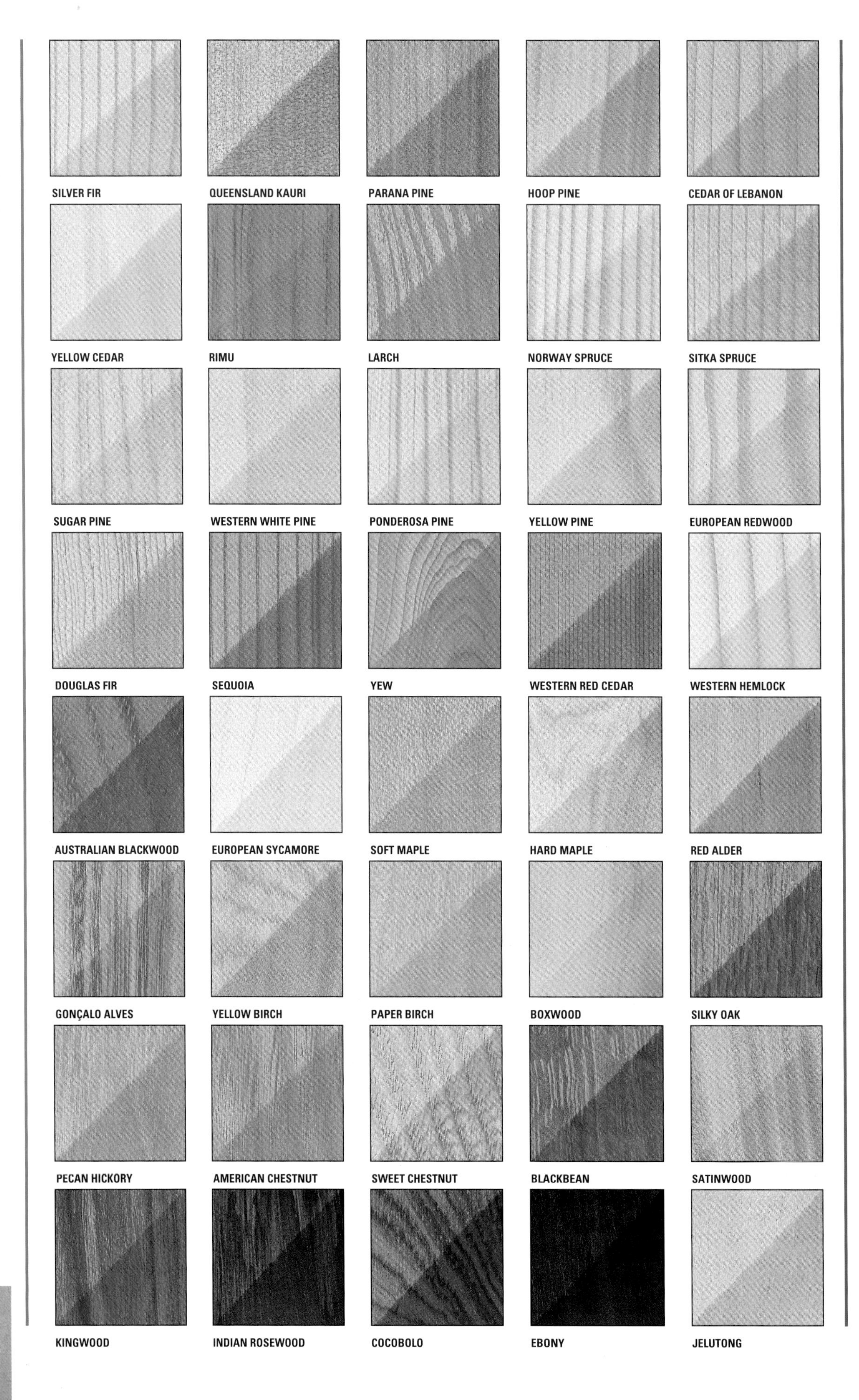
SILVER FIR
QUEENSLAND KAURI
PARANA PINE
HOOP PINE
CEDAR OF LEBANON
YELLOW CEDAR
RIMU
LARCH
NORWAY SPRUCE
SITKA SPRUCE
SUGAR PINE
WESTERN WHITE PINE
PONDEROSA PINE
YELLOW PINE
EUROPEAN REDWOOD
DOUGLAS FIR
SEQUOIA
YEW
WESTERN RED CEDAR
WESTERN HEMLOCK
AUSTRALIAN BLACKWOOD
EUROPEAN SYCAMORE
SOFT MAPLE
HARD MAPLE
RED ALDER
GONÇALO ALVES
YELLOW BIRCH
PAPER BIRCH
BOXWOOD
SILKY OAK
PECAN HICKORY
AMERICAN CHESTNUT
SWEET CHESTNUT
BLACKBEAN
SATINWOOD
KINGWOOD
INDIAN ROSEWOOD
COCOBOLO
EBONY
JELUTONG

WOOD FINISHES

It is the nature of wood to be as varied in its colour as in its figure and texture. Even when prepared and finished, wood will continue to respond to its environment; not only will it 'move', but the colour will also alter in time by becoming lighter or darker, according to species. The result of this process is known as patina.

Colour change

The most dramatic changes in colour occur when a finish is applied; even a clear finish enriches and slightly darkens natural colours. The woods illustrated on pages 44–53 and 56–82 are shown here for comparison in actual-size samples, illustrating the wood before and after the application of a clear finish.

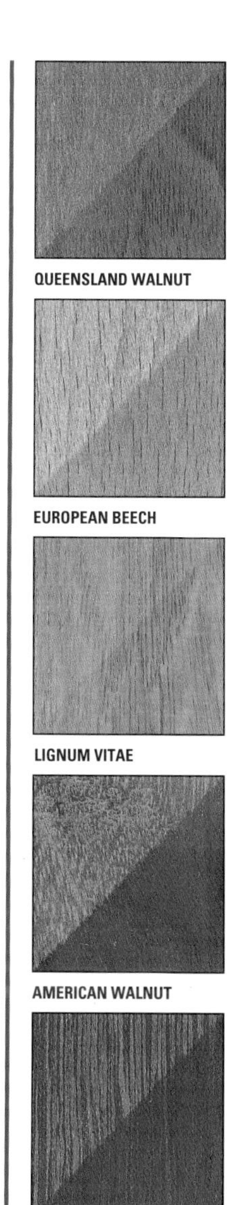

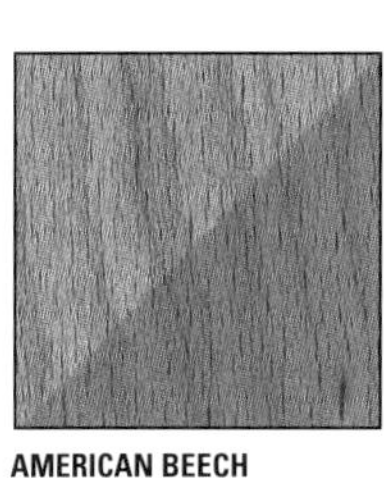

QUEENSLAND WALNUT

UTILE

JARRAH

AMERICAN BEECH

EUROPEAN BEECH

AMERICAN WHITE ASH

EUROPEAN ASH

RAMIN

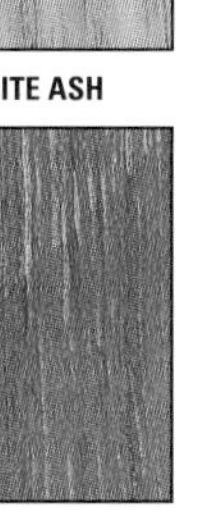

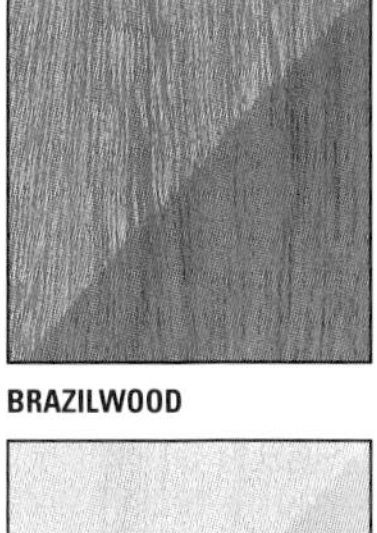

LIGNUM VITAE

BUBINGA

BRAZILWOOD

BUTTERNUT

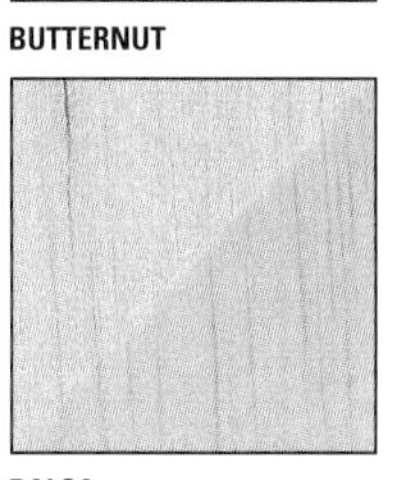

AMERICAN WALNUT

EUROPEAN WALNUT

AMERICAN WHITEWOOD

BALSA

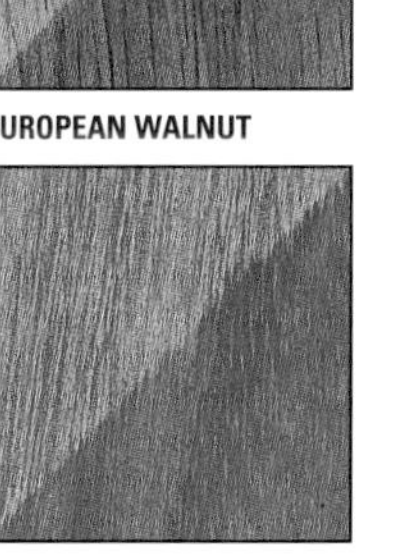

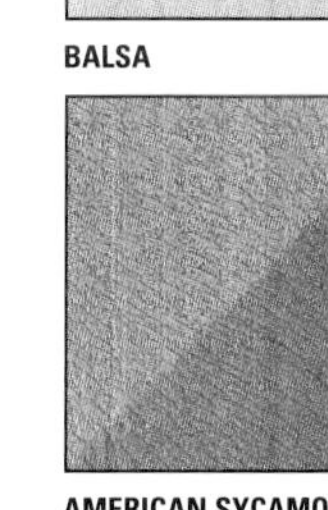

PURPLEHEART

AFRORMOSIA

EUROPEAN PLANE

AMERICAN SYCAMORE

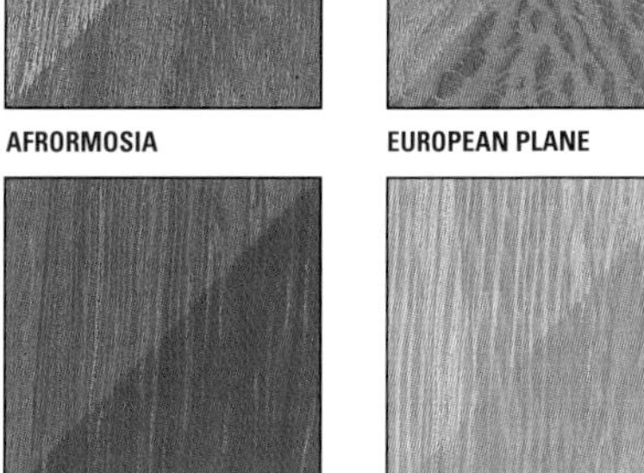

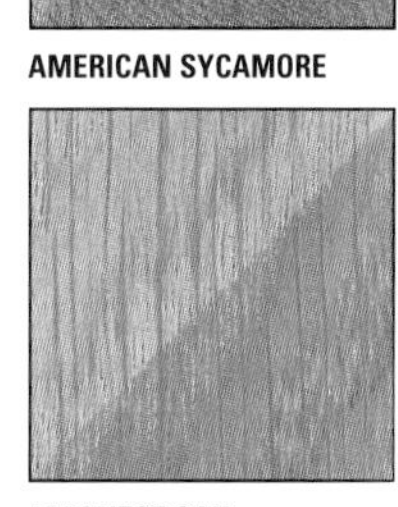

AMERICAN CHERRY

AFRICAN PADAUK

AMERICAN WHITE OAK

JAPANESE OAK

AMERICAN RED OAK

RED LAUAN

BRAZILIAN MAHOGANY

TEAK

EUROPEAN OAK

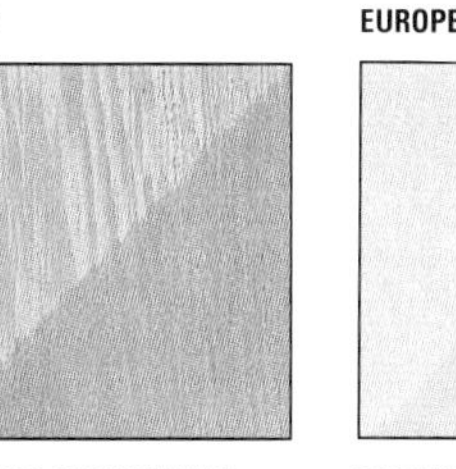

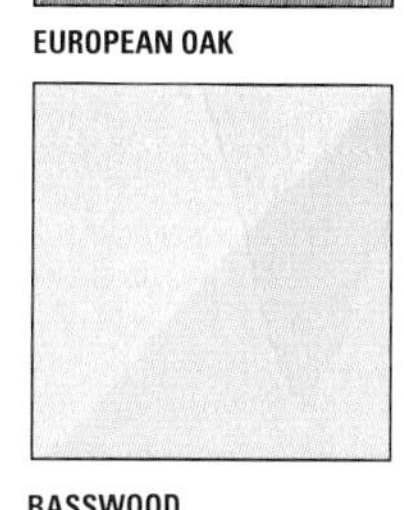

LIME

OBECHE

AMERICAN WHITE ELM

DUTCH & ENGLISH ELM

BASSWOOD

TREE SPECIES

The qualities of wood – figure, colour, workability and even scent – are usually the features of most interest to the wood-worker. However, it is hard to imagine that anyone who works with wood will not be fascinated by the trees from which such a versatile material comes. Set out here are illustrations of the trees from which the woods shown on pages 44–53 and 56–82 are harvested. These are not to scale, but represent the common form taken by the trees when growing in an ideal natural environment.

Kingwood

Cocobolo

Ramin

Botanical details
The authors and editors have consulted botanical libraries and known reference sources in an attempt to illustrate a profile of each tree featured in this book. Unfortunately, no known visual reference of the whole shape and configuration of the above trees could be located, and these have been substituted with illustrations of the available botanical details.

Softwoods
from left to right:

(top row)	*(second row)*	*(third row)*	*(bottom row)*
Silver fir	Yellow cedar	Sugar pine	Douglas fir
Queensland kauri	Rimu	Western white pine	Sequoia
Parana pine	Larch	Ponderosa pine	Yew
Hoop pine	Norway spruce	Yellow pine	Western red cedar
Cedar of Lebanon	Sitka spruce	European redwood	Western hemlock

Hardwoods
from left to right, in sequence:
Australian blackwood
European sycamore
Soft maple
Hard maple
Red alder
Gonçalo alves
Yellow birch
Paper birch
Boxwood
Silky oak
Pecan hickory
American chestnut
Sweet chestnut
Blackbean
Satinwood
Indian rosewood
Ebony
Jelutong
Queensland walnut
Utile
Jarrah
American beech
European beech
American white ash
European ash
Lignum vitae
Bubinga
Brazilwood
Butternut

Hardwoods
(continued)
from left to right, in sequence:
American walnut
European walnut
American whitewood
Balsa
Purpleheart
Afrormosia
European plane
American sycamore
American cherry
African padauk
American white oak
Japanese oak
European oak
American red oak
Red lauan
Brazilian mahogany
Teak
Basswood
Lime
Obeche
American white elm
Dutch & English elm

GLOSSARY OF TERMS

Air-drying
A method for seasoning timber that permits covered stacks of sawn wood to dry naturally in the open air.

Alburnum
Another name for sapwood.

Animal glue
A protein-based wood glue made from animal skins and bone.

Autoclave
A sealed pressure vessel used in the production of dyed veneer.

B

Backing grade
The category of cheaper veneers that are glued to the back of a board in order to balance better-quality veneers glued to the front face.

Banding
A plain or patterned strip of veneer used to make decorative borders.

Batten
A strip of wood.

Bevel
A surface that meets another at an angle other than a right angle. *or* To cut such surfaces.

Blank
A piece of wood roughly cut to size ready for turning on a lathe.

Bleeding
The process in which a natural substance, for instance wood resin, permeates and stains the surface of a subsequent covering or finish.

Blemish
Any defect that mars the appearance of wood.

Blister
A small raised area of veneer resulting from insufficient glue at that point.

Blockboard
A man-made building board with a core of approximately square-section solid-wood strips sandwiched between thin plywood sheets. *See also* laminboard.

Bowed
Twisted wood.

Bruise
To make a dent in timber by striking it with a hard object such as a hammer.

Burl
See burr.

Burr
A warty growth on a tree trunk. When sliced, it produces speckled burr veneer. *or* A very thin strip of metal left along the cutting edge of a blade after honing or grinding.

Buttressed
Having roughly triangular outgrowths at the foot of a tree trunk to provide increased stability.

C

Case-hardened
A term used to describe unevenly seasoned timber with a moisture content that varies throughout its thickness.

Cauls
Sheets of wood or metal used to press veneer onto groundwork. They can be flat or curved as required.

Checks
Splits in timber caused by uneven seasoning. *See also* knife checks.

Chipboard
A mat of small particles of wood and glue compressed into a flat building board.

Clear timber
Good-quality wood that is free from defects.

Close grain
A term used to describe wood with small pores or fine cell structure. Also known as fine-textured.

Coarse-textured
See open grain.

Comb-grain
Another term for quarter-sawn.

Common (COM)
An American hardwood grading term for wood below FAS and Selects. There are four subdivisions.

Contact glue
An adhesive that bonds to itself without the aid of cramps when two previously glued surfaces are brought together.

Concave
Curving inwards.

Convex
Curving outwards.

Core
The central layer of plies, particles or wooden strips in a man-made board.

Cross-banding
Strips of veneer cut across the grain and used as decorative borders.

Cross-bearers
Lengths of wood used to provide pressure on cauls.

Crosscutting
Sawing across the grain.

Cross grain
Grain that deviates from the main axis of a workpiece or tree.

Cross ply
A veneer laid perpendicular beneath the face veneer of laminated boards.

Crotch figure
Another term for curl figure.

Crown-cut
A term used to describe veneer that has been tangentially sliced from a log, producing oval or curved grain patterns.

Cup
To bend as a result of shrinkage – specifically across the width of a piece of wood.

Cure
To set as a result of chemical reaction.

Curl figure
The grain pattern on wood that has been cut from that part of a tree where a branch joins the main stem or trunk.

Curly grain
Wood grain exhibiting an irregular wavy pattern.

Defect
Any abnormality or irregularity that decreases wood's working properties and value.

Diffuse-porous
In hardwoods, where the pores are roughly the same size in both the sapwood and heartwood. *See also* ring-porous.

Dimension stock
Prepared timber cut to standard sizes.

Dressed stock
Another term for dimension stock.

Durability
Resistance to deterioration, particularly from decay.

Earlywood
That part of a tree's annual growth ring that is laid down in the early part of the growing season.

Edge grain
Another term for quarter-sawn.

End grain
The surface of wood exposed after cutting across the fibres.

EMC
Equilibrium moisture content – the moisture content reached by a piece of timber when exposed to a more or less constant level of temperature and humidity.

F

Face edge
The surface planed square to the face side, and from which other dimensions and angles are measured.

Face quality
A term used to describe better-quality veneers that are used to cover the visible surfaces of a workpiece.

Face side
The flat planed surface from which all other dimensions and angles are measured.

FAS
'Firsts and seconds'. The best American grade for commercial hardwoods.

Fibreboards
A range of building boards made from reconstituted wood fibres.

Figure
Another term for grain pattern.

Fillet
A narrow strip of timber.

Fine-textured
See close grain.

Flat-grain
Another term for plain-sawn.

Flat-sliced
A term used to describe a narrow sheet of veneer cut from part of a log with a knife, using a slicing action.

Flat-sawn
Another term for plain-sawn.

Flitches
Pieces of wood that are sawn from a log for slicing into veneers. *or* The bundle of sliced veneers.

FS
'Fresh-sawn'. Wood supplied newly cut from a log.

G

Grade
A term used to define the quality of a log or sawn wood.

Grain
The general direction or arrangement of the fibrous materials of wood.

Green wood
Newly cut timber that has not been seasoned.

Groundwork
The backing material to which veneer is glued.

H

Hardwood
Wood cut from broadleaved, mostly deciduous, trees that belong to the botanical group *Angiospermae*.

Heartwood
Mature wood that forms the spine of a tree.

I

Inlay
To insert pieces of wood or metal into prepared recesses so that the material lies flush with the surrounding surfaces. *or* The piece of material itself.

Interlocked grain
Bands of annual-growth rings with alternating right-hand and left-hand spiral grain.

In wind
See winding.

K

KD
Timber trade term for kiln-dried wood.

Kerf
The slot cut by a saw.

Key
To abrade or incise a surface to provide a good grip for gluing.

Kiln-drying
A method for seasoning timber, using a mixture of hot air and steam.

Knife checks
Splits across veneer caused by poorly adjusted veneer-slicing equipment.

Knotting
A shellac-based sealer used to coat resinous knots that would otherwise stain subsequent finishes. *See also* bleeding.

L

Laminate
A component made from thin strips of wood glued together. *or* To glue strips together to form a component.

Laminboard
A man-made building board with a core of narrow strips of wood glued together and sandwiched between thin plywood sheets. *See also* blockboard.

Latewood
That part of a tree's annual-growth ring that is laid down in the latter part of the growing season.

Lipping
A protective strip of solid wood that is applied to the edge of a man-made-board panel or table top.

Long grain
Grain that is aligned with the main axis of a workpiece. *See also* short grain.

M

Marquetry
The process of laying relatively small pieces of veneer to make decorative patterns or pictures. *See also* parquetry.

Mild
Easy to work.

Mitre
A joint formed between two pieces of wood by cutting bevels of equal angles (usually 45 degrees) at the ends of both pieces. *or* To cut such a joint.

Moisture content
The proportion by weight of water present in the tissues of a piece of timber, given as a percentage of the oven-dry weight. Also known as MC in the timber trade.

N

Nominal dimensions
Standardized widths and thicknesses of timber newly sawn from a log. The actual sizes are reduced by shrinkage and planing.

O

Open grain
A term used to describe ring-porous wood with large pores. Also known as coarse-textured.

P

PAR
'Planed all round'. Wood with all its sides planed after dimension sawing.

Parquetry
A similar process to marquetry, but using veneers cut into geometric shapes to make decorative patterns.

Particle boards
Building boards made from small chips or flakes of wood bonded together with glue under pressure.

Patina
The colour and texture that a material such as wood or metal acquires as a result of a natural ageing process.

PBS
'Planed both sides'. Wood with the front and back faces planed after dimension sawing.

PEG
Polyethylene glycol – a stabilizing agent used in place of conventional seasoning processes to treat green timber.

Photosynthethis
A natural process that takes place when energy in the form of light is absorbed by chlorophyll, producing the nutrients on which plants live.

Plain-sawn
A term used to describe a piece of wood with growth rings that meet the faces of the board at angles of less than 45 degrees. *See also* rift-sawn.

Plywood
A building board made by bonding a number of wood veneers together under pressure.

PS
'Part-seasoned'. Some dense woods are difficult to season, and are sold as PS boards, with no guarantee as to the moisture content.

Q

Quarter-sawn
A term used to describe a piece of wood with growth rings at not less than 45 degrees to the faces of the board. *See also* rift-sawn.

R

Rift-sawn
A term used to describe a piece of wood with growth rings that meet the faces of the board at angles of more than 30 degrees, but at less than 60 degrees.

Ring-porous
In hardwoods, where larger pores are found in earlywood and smaller ones in latewood. *See also* diffuse-porous.

Ripsawing
Cutting along the grain.

Rotary-cut
A term used to describe a continuous sheet of veneer peeled from a log by turning it against a stationary knife.

S

Sapwood
New wood surrounding the denser heartwood.

SE (Squares)
Boards that are cut square on both edges.

Seasoning
Reducing the moisture content of timber.

Selects (SEL)
The second-best American grade for hardwoods.

Short grain
A term used to describe where the general direction of wood fibres lies across a narrow section of timber.

Slash-sawn
Another term for plain-sawn.

Softwood
Wood cut from coniferous trees which belong to the botanical group *Gymnospermae*.

Sound
Wood that has no decay.

Spalted
Partially decayed wood with irregular discoloration bounded by dark 'zone lines'.

Springwood
Another term for earlywood.

Straight grain
Grain that aligns with the main axis of a workpiece or tree.

Stringing lines
Fine strips of wood used to divide areas of veneer.

Summerwood
Another term for latewood.

T

Tangentially cut
Another term for plain-sawn.

Thermoplastic
A term used to describe a material that can be resoftened with heat, such as animal glue.

Thermosetting
A term used to describe a material that cannot be resoftened with heat once it has set hard, such as resin glue.

V

Veneer
A thin slice of wood used as a surface covering on a less expensive material, such as a man-made board.

Vertical grain
Another term for quarter-sawn.

W

Waney edge
The natural wavy edge of a plank, that may still be covered by tree bark.

Wavy grain
A term used to describe the even, wave-like grain pattern on wood cut from a tree with an undulating cell structure.

Wild grain
Irregular grain that changes direction, making it difficult to work.

Winding
A warped or twisted board is said to be winding or in wind.

Z

Zone lines
See spalted.

A

B

C

D

E

F